IET ENERGY ENGINEERING SERIES 235

Clean Energy

Other volumes in this series:

Volume 1	**Power Circuit Breaker Theory and Design** C.H. Flurscheim (Editor)
Volume 4	**Industrial Microwave Heating** A.C. Metaxas and R.J. Meredith
Volume 7	**Insulators for High Voltages** J.S.T. Looms
Volume 8	**Variable Frequency AC Motor Drive Systems** D. Finney
Volume 10	**SF_6 Switchgear** H.M. Ryan and G.R. Jones
Volume 11	**Conduction and Induction Heating** E.J. Davies
Volume 13	**Statistical Techniques for High Voltage Engineering** W. Hauschild and W. Mosch
Volume 14	**Uninterruptible Power Supplies** J. Platts and J.D. St Aubyn (Editors)
Volume 15	**Digital Protection for Power Systems** A.T. Johns and S.K. Salman
Volume 16	**Electricity Economics and Planning** T.W. Berrie
Volume 18	**Vacuum Switchgear** A. Greenwood
Volume 19	**Electrical Safety: A guide to causes and prevention of hazards** J. Maxwell Adams
Volume 21	**Electricity Distribution Network Design, 2nd Edition** E. Lakervi and E.J. Holmes
Volume 22	**Artificial Intelligence Techniques in Power Systems** K. Warwick, A.O. Ekwue and R. Aggarwal (Editors)
Volume 24	**Power System Commissioning and Maintenance Practice** K. Harker
Volume 25	**Engineers' Handbook of Industrial Microwave Heating** R.J. Meredith
Volume 26	**Small Electric Motors** H. Moczala *et al.*
Volume 27	**AC–DC Power System Analysis** J. Arrillaga and B.C. Smith
Volume 29	**High Voltage Direct Current Transmission, 2nd Edition** J. Arrillaga
Volume 30	**Flexible AC Transmission Systems (FACTS)** Y.-H. Song (Editor)
Volume 31	**Embedded Generation** N. Jenkins *et al.*
Volume 32	**High Voltage Engineering and Testing, 2nd Edition** H.M. Ryan (Editor)
Volume 33	**Overvoltage Protection of Low-Voltage Systems, Revised Edition** P. Hasse
Volume 36	**Voltage Quality in Electrical Power Systems** J. Schlabbach *et al.*
Volume 37	**Electrical Steels for Rotating Machines** P. Beckley
Volume 38	**The Electric Car: Development and future of battery, hybrid and fuel-cell cars** M. Westbrook
Volume 39	**Power Systems Electromagnetic Transients Simulation** J. Arrillaga and N. Watson
Volume 40	**Advances in High Voltage Engineering** M. Haddad and D. Warne
Volume 41	**Electrical Operation of Electrostatic Precipitators** K. Parker
Volume 43	**Thermal Power Plant Simulation and Control** D. Flynn
Volume 44	**Economic Evaluation of Projects in the Electricity Supply Industry** H. Khatib
Volume 45	**Propulsion Systems for Hybrid Vehicles** J. Miller
Volume 46	**Distribution Switchgear** S. Stewart
Volume 47	**Protection of Electricity Distribution Networks, 2nd Edition** J. Gers and E. Holmes
Volume 48	**Wood Pole Overhead Lines** B. Wareing
Volume 49	**Electric Fuses, 3rd Edition** A. Wright and G. Newbery
Volume 50	**Wind Power Integration: Connection and system operational aspects** B. Fox *et al.*
Volume 51	**Short Circuit Currents** J. Schlabbach
Volume 52	**Nuclear Power** J. Wood
Volume 53	**Condition Assessment of High Voltage Insulation in Power System Equipment** R.E. James and Q. Su
Volume 55	**Local Energy: Distributed generation of heat and power** J. Wood
Volume 56	**Condition Monitoring of Rotating Electrical Machines** P. Tavner, L. Ran, J. Penman and H. Sedding
Volume 57	**The Control Techniques Drives and Controls Handbook, 2nd Edition** B. Drury
Volume 58	**Lightning Protection** V. Cooray (Editor)
Volume 59	**Ultracapacitor Applications** J.M. Miller
Volume 62	**Lightning Electromagnetics** V. Cooray

Volume 63 **Energy Storage for Power Systems, 2nd Edition** A. Ter-Gazarian
Volume 65 **Protection of Electricity Distribution Networks, 3rd Edition** J. Gers
Volume 66 **High Voltage Engineering Testing, 3rd Edition** H. Ryan (Editor)
Volume 67 **Multicore Simulation of Power System Transients** F.M. Uriate
Volume 68 **Distribution System Analysis and Automation** J. Gers
Volume 69 **The Lightening Flash, 2nd Edition** V. Cooray (Editor)
Volume 70 **Economic Evaluation of Projects in the Electricity Supply Industry, 3rd Edition** H. Khatib
Volume 72 **Control Circuits in Power Electronics: Practical issues in design and implementation** M. Castilla (Editor)
Volume 73 **Wide Area Monitoring, Protection and Control Systems: The enabler for smarter grids** A. Vaccaro and A. Zobaa (Editors)
Volume 74 **Power Electronic Converters and Systems: Frontiers and applications** A.M. Trzynadlowski (Editor)
Volume 75 **Power Distribution Automation** B. Das (Editor)
Volume 76 **Power System Stability: Modelling, analysis and control** A.A. Sallam and B. Om P. Malik
Volume 78 **Numerical Analysis of Power System Transients and Dynamics** A. Ametani (Editor)
Volume 79 **Vehicle-to-Grid: Linking electric vehicles to the smart grid** J. Lu and J. Hossain (Editors)
Volume 81 **Cyber-Physical-Social Systems and Constructs in Electric Power Engineering** S. Suryanarayanan, R. Roche and T.M. Hansen (Editors)
Volume 82 **Periodic Control of Power Electronic Converters** F. Blaabjerg, K. Zhou, D. Wang and Y. Yang
Volume 86 **Advances in Power System Modelling, Control and Stability Analysis** F. Milano (Editor)
Volume 87 **Cogeneration: Technologies, optimisation and implementation** C.A. Frangopoulos (Editor)
Volume 88 **Smarter Energy: From smart metering to the smart grid** H. Sun, N. Hatziargyriou, H.V. Poor, L. Carpanini and M.A. Sánchez Fornié (Editors)
Volume 89 **Hydrogen Production, Separation and Purification for Energy** A. Basile, F. Dalena, J. Tong and T.N. Veziroğlu (Editors)
Volume 90 **Clean Energy Microgrids** S. Obara and J. Morel (Editors)
Volume 91 **Fuzzy Logic Control in Energy Systems with Design Applications in MATLAB®/Simulink®** İ.H. Altaş
Volume 92 **Power Quality in Future Electrical Power Systems** A.F. Zobaa and S.H.E.A. Aleem (Editors)
Volume 93 **Cogeneration and District Energy Systems: Modelling, analysis and optimization** M.A. Rosen and S. Koohi-Fayegh
Volume 94 **Introduction to the Smart Grid: Concepts, technologies and evolution** S.K. Salman
Volume 95 **Communication, Control and Security Challenges for the Smart Grid** S.M. Muyeen and S. Rahman (Editors)
Volume 96 **Industrial Power Systems with Distributed and Embedded Generation** R Belu
Volume 97 **Synchronized Phasor Measurements for Smart Grids** M.J.B. Reddy and D.K. Mohanta (Editors)
Volume 98 **Large Scale Grid Integration of Renewable Energy Sources** A. Moreno-Munoz (Editor)
Volume 100 **Modeling and Dynamic Behaviour of Hydropower Plants** N. Kishor and J. Fraile-Ardanuy (Editors)
Volume 101 **Methane and Hydrogen for Energy Storage** R. Carriveau and D.S.-K. Ting
Volume 104 **Power Transformer Condition Monitoring and Diagnosis** A. Abu-Siada (Editor)
Volume 106 **Surface Passivation of Industrial Crystalline Silicon Solar Cells** J. John (Editor)
Volume 107 **Bifacial Photovoltaics: Technology, applications and economics** J. Libal and R. Kopecek (Editors)

Volume 108 **Fault Diagnosis of Induction Motors** J. Faiz, V. Ghorbanian and G. Joksimović
Volume 109 **Cooling of Rotating Electrical Machines: Fundamentals, modelling, testing and design** D. Staton, E. Chong, S. Pickering and A. Boglietti
Volume 110 **High Voltage Power Network Construction** K. Harker
Volume 111 **Energy Storage at Different Voltage Levels: Technology, integration, and market aspects** A.F. Zobaa, P.F. Ribeiro, S.H.A. Aleem and S.N. Afifi (Editors)
Volume 112 **Wireless Power Transfer: Theory, technology and application** N. Shinohara
Volume 114 **Lightning-Induced Effects in Electrical and Telecommunication Systems** Y. Baba and V.A. Rakov
Volume 115 **DC Distribution Systems and Microgrids** T. Dragičević, F. Blaabjerg and P. Wheeler
Volume 116 **Modelling and Simulation of HVDC Transmission** M. Han (Editor)
Volume 117 **Structural Control and Fault Detection of Wind Turbine Systems** H.R. Karimi
Volume 118 **Modelling and Simulation of Complex Power Systems** A. Monti and A. Benigni
Volume 119 **Thermal Power Plant Control and Instrumentation: The control of boilers and HRSGs, 2nd Edition** D. Lindsley, J. Grist and D. Parker
Volume 120 **Fault Diagnosis for Robust Inverter Power Drives** A. Ginart (Editor)
Volume 121 **Monitoring and Control using Synchrophasors in Power Systems with Renewables** I. Kamwa and C. Lu (Editors)
Volume 123 **Power Systems Electromagnetic Transients Simulation, 2nd Edition** N. Watson and J. Arrillaga
Volume 124 **Power Market Transformation** B. Murray
Volume 125 **Wind Energy Modeling and Simulation, Volume 1: Atmosphere and plant** P. Veers (Editor)
Volume 126 **Diagnosis and Fault Tolerance of Electrical Machines, Power Electronics and Drives** A.J.M. Cardoso
Volume 127 **Lightning Electromagnetics, Volume 1: Return stroke modelling and electromagnetic radiation, 2nd Edition** V. Cooray, F. Rachidi, M. Rubinstein (Editors)
Volume 127 **Lightning Electromagnetics, Volume 2: Electrical processes and effects, 2nd Edition** V. Cooray, F. Rachidi, M. Rubinstein (Editors)
Volume 128 **Characterization of Wide Bandgap Power Semiconductor Devices** F. Wang, Z. Zhang and E.A. Jones
Volume 129 **Renewable Energy from the Oceans: From wave, tidal and gradient systems to offshore wind and solar** D. Coiro and T. Sant (Editors)
Volume 130 **Wind and Solar Based Energy Systems for Communities** R. Carriveau and D.S.-K. Ting (Editors)
Volume 131 **Metaheuristic Optimization in Power Engineering** J. Radosavljević
Volume 132 **Power Line Communication Systems for Smart Grids** I.R.S. Casella and A. Anpalagan
Volume 134 **Hydrogen Passivation and Laser Doping for Silicon Solar Cells** B. Hallam and C. Chan (Editors)
Volume 139 **Variability, Scalability and Stability of Microgrids** S.M. Muyeen, S.M. Islam and F. Blaabjerg (Editors)
Volume 142A **Wind Turbine System Design, Volume 1: Nacelles, drive trains and verification** J. Wenske (Editor)
Volume 142B **Wind Turbine System Design, Volume 2: Electrical systems, grid integration, control and monitoring** J. Wenske (Editor)
Volume 143 **Medium Voltage DC System Architectures** B. Grainger and R.D. Doncker (Editors)
Volume 145 **Condition Monitoring of Rotating Electrical Machines** P. Tavner, L. Ran, and C. Crabtree
Volume 146 **Energy Storage for Power Systems, 3rd Edition** A.G. Ter-Gazarian
Volume 147 **Distribution Systems Analysis and Automation, 2nd Edition** J. Gers
Volume 151 **SiC Power Module Design: Performance, robustness and reliability** A. Castellazzi and A. Irace (Editors)

Volume 152 **Power Electronic Devices: Applications, failure mechanisms and reliability** F Iannuzzo (Editor)
Volume 153 **Signal Processing for Fault Detection and Diagnosis in Electric Machines and Systems** M. Benbouzid (Editor)
Volume 155 **Energy Generation and Efficiency Technologies for Green Residential Buildings** D. Ting and R. Carriveau (Editors)
Volume 156 **Lithium-ion Batteries Enabled by Silicon Anodes** C. Ban and K. Xu (Editors)
Volume 157 **Electrical Steels, 2 Volumes** A. Moses, K. Jenkins, Philip Anderson and H. Stanbury
Volume 158 **Advanced Dielectric Materials for Electrostatic Capacitors** Q. Li (Editor)
Volume 159 **Transforming the Grid Towards Fully Renewable Energy** O. Probst, S. Castellanos and R. Palacios (Editors)
Volume 160 **Microgrids for Rural Areas: Research and case studies** R.K. Chauhan, K. Chauhan and S.N. Singh (Editors)
Volume 161 **Artificial Intelligence for Smarter Power Systems: Fuzzy logic and neural networks** M.G. Simoes
Volume 165 **Digital Protection for Power Systems, 2nd Edition** S.K. Salman
Volume 166 **Advanced Characterization of Thin Film Solar Cells** N. Haegel and M. Al-Jassim (Editors)
Volume 167 **Power Grids with Renewable Energy: Storage, integration and digitalization** A.A. Sallam and B. OM P. Malik
Volume 169 **Small Wind and Hydrokinetic Turbines** P. Clausen, J. Whale and D. Wood (Editors)
Volume 170 **Reliability of Power Electronics Converters for Solar Photovoltaic Applications** F. Blaabjerg, A.L. Haque, H. Wang, Z. Abdin Jaffery and Y. Yang (Editors)
Volume 171 **Utility-Scale Wind Turbines and Wind Farms** A. Vasel-Be-Hagh and D.S.-K. Ting
Volume 172 **Lighting Interaction with Power Systems, 2 Volumes** A. Piantini (Editor)
Volume 174 **Silicon Solar Cell Metallization and Module Technology** T. Dullweber (Editor)
Volume 175 **n-Type Crystalline Silicon Photovoltaics: Technology, applications and economics** D. Munoz and R. Kopecek (Editors)
Volume 178 **Integrated Motor Drives** X. Deng and B. Mecrow (Editors)
Volume 180 **Protection of Electricity Distribution Networks, 4th Edition** J. Gers and E. Holmes
Volume 182 **Surge Protection for Low Voltage Systems** A. Rousseau (Editor)
Volume 184 **Compressed Air Energy Storage: Types, systems and applications** D. Ting and J. Stagner
Volume 186 **Synchronous Reluctance Machines: Analysis, optimization and applications** N. Bianchi, C. Babetto and G. Bacco
Volume 190 **Synchrophasor Technology: Real-time operation of power networks** N. Kishor and S.R. Mohanty (Editors)
Volume 191 **Electric Fuses: Fundamentals and new applications, 4th Edition** N. Nurse, A. Wright and P.G. Newbery
Volume 193 **Overhead Electric Power Lines: Theory and practice** S. Chattopadhyay and A. Das
Volume 194 **Offshore Wind Power: Reliability, availability and maintenance, 2nd edition** P. Tavner
Volume 196 **Cyber Security for Microgrids** S. Sahoo, F. Blaajberg and T. Dragicevic
Volume 198 **Battery Management Systems and Inductive Balancing** A. Van den Bossche and A. Farzan Moghaddam
Volume 199 **Model Predictive Control for Microgrids: From power electronic converters to energy management** J. Hu, J.M. Guerrero and S. Islam
Volume 201 **Design, Control and Monitoring of Tidal Stream Turbine Systems** M. Benbouzid (Editor)
Volume 204 **Electromagnetic Transients in Large HV Cable Networks: Modeling and calculations** Ametani, Xue, Ohno and Khalilnezhad

Volume 208 **Nanogrids and Picogrids and their Integration with Electric Vehicles** S. Chattopadhyay
Volume 210 **Superconducting Magnetic Energy Storage in Power Grids** M.H. Ali
Volume 211 **Blockchain Technology for Smart Grids: Implementation, management and security** H.L. Gururaj, K.V. Ravi, F. Flammini, H. Lin, B. Goutham, K.B.R. Sunil and C. Sivapragash
Volume 212 **Battery State Estimation: Methods and models** S. Wang
Volume 215 **Industrial Demand Response: Methods, best practices, case studies, and applications** H.H. Alhelou, A. Moreno-Muñoz and P. Siano (Editors)
Volume 213 **Wide Area Monitoring of Interconnected Power Systems, 2nd Edition** A.R. Messina
Volume 214 **Distributed Energy Storage in Urban Smart Grids** P.F. Ribeiro and R.S. Salles (Editors)
Volume 217 **Advances in Power System Modelling, Control and Stability Analysis, 2nd Edition** F. Milano (Editor)
Volume 225 **Fusion-Fission Hybrid Nuclear Reactors: For enhanced nuclear fuel utilization and radioactive waste reduction** W.M. Stacey
Volume 227 **Processing and Manufacturing of Electrodes for Lithium-Ion Batteries** J. Li and C. Jin (Editors)
Volume 228 **Digital Technologies for Solar Photovoltaic Systems: From general to rural and remote installations** S. Motahhir (Editor)
Volume 236 **Green HV Switching Technologies for Modern Power Networks,** K. Niayesh (Editor)
Volume 238 **AI for Status Monitoring of Utility Scale Batteries** S. Wang, K. Liu, Y. Wang, D. Stroe, C. Fernandez and J.M. Guerrero
Volume 239 **Intelligent Control of Medium and High Power Converters** M. Bendaoud, Y. Maleh and S. Padmanaban (Editors)
Volume 247 **Power System Strength: Evaluation methods, best practice, case studies, and applications** H. Alhelou, N. Hosseinzadeh and B. Bahrani (Editors)
Volume 905 **Power System Protection, 4 Volumes**

Clean Energy

Past to future

Peter Tavner

The Institution of Engineering and Technology

Published by The Institution of Engineering and Technology, London, United Kingdom

The Institution of Engineering and Technology is registered as a Charity in England & Wales (no. 211014) and Scotland (no. SC038698).

First published 2023

The Institution of Engineering and Technology
Futures Place
Kings Way, Stevenage
Hertfordshire, SG1 2UA, United Kingdom

www.theiet.org

British Library Cataloguing in Publication Data
A catalogue record for this product is available from the British Library

ISBN 978-1-83953-712-7 (hardback)
ISBN 978-1-83953-713-4 (PDF)

Typeset in India by MPS Limited

Cover image: Yaorusheng/Moment via Getty Images

This book is dedicated to:

Elizabeth Helen Tavner, née Moore

A Poem:

I am the song that sings the bird. I am the leaf that grows the land.
I am the tide that moves the moon. I am the stream that halts the sand.
I am the cloud that drives the storm. I am the earth that lights the sun.
I am the fire that strikes the stone. I am the clay that shapes the hand.
I am the word that speaks the man.
Charles Causley (1986)

Music: The First Morning of the World.
Rachel Portman (2021)

Contents

About the author

Peter Tavner, Eur Ing, CEng, FIET is an Emeritus Professor of Durham University, UK with a Cambridge MA 1972, a C A Parsons Ltd sponsored Southampton PhD 1978, and a Durham DSc 2012.

He served as a Royal Navy officer in General Service on HM Ships ASHANTI, BRISTOL and LONDON, subsequently working in a Medical Research Council laboratory and teaching at Benin University, Nigeria.

From 1979 to 2002, he held research and management positions in the electrical supply and manufacturing industries, including Section Head in a former CEGB laboratory, Technical Director of Laurence, Scott & Electromotors Ltd (LSE) then of Brush Electrical Machines Ltd, two of UK's electrical machine manufacturers, finally Group Technical Director of FKI Energy Technology, an international business, owner of LSE and Brush, manufacturing electrical machines, drives and wind turbines, with factories in the UK, Holland, Italy, Germany and the Czech Republic.

In 2003, he became Professor of New & Renewable Energy at Durham University, Department of Engineering, leading several UK and European research initiatives to improve wind, wave and tidal reliability, becoming Head of Engineering from 2006 to 2010. More recently he has provided consultancy to his former company LSE in UK and to Dong Fang Electrical Machinery (DFEM) in Deyang, Szechuan, China, he has also taught at Chongqing University (CQU), China.

He retains an interest in practical electrical machine analysis, the improvement of power equipment reliability and especially the development of wind, tidal and wave power.

In 1990, he won an IEE Institution Premium with Richard Jackson for a paper on discharge current coupling between electrical machine conductors, in 2020, he won the European Academy of Wind Energy Scientific Award for a pioneering role in research into wind farm reliability, availability and maintenance. He was non-exec Director of UK's New and Renewable Energy Centre (NaREC), Blyth, now ORE Catapult, is an Honorary Member of the European Academy of Wind Energy and has been Scientific Advisory Board Chair to the following UK Research and Innovation consortia:

- Centre for Power Electronics, led by Nottingham University;
- HOME Offshore Wind Farms Project, led by Manchester University;
- New Partnership for Offshore Wind (NPOW) between Sheffield, Durham and Hull Universities supported by Siemens-Gamesa and Ørsted;
- AURA Centre for NPOW Doctoral Training, led by Hull University.

Preface

This book is aimed at those with an interest in energy, history and the mitigation of climate change. At a time of increasing human concern about the climate, it describes the scale of human methods for extracting energy from our environment and their effects upon it, up to and including the modern age.

The author is not a historian but an engineer, having worked in the engine rooms of steam and gas turbine warships, the machinery halls of large power stations, in the electrical generator, power electronic and wind farm manufacturing industries and subsequently on the study and teaching of wind, wave and tidal power.

The object of this book is not to be a scientific treatise but to tell the human, engineering and scientific story of clean energy development to date, how we understand energy, how we use our technology and how we plan to use it to reduce its impact on our environment.

Both technical and non-technical readers should find the book valuable for understanding our present energy predicament, in context, with clean energy solutions to resolve it.

This is a difficult task, the author will guide the reader to consider first the Earth's ancient origin and species development, then turn particularly to plant, animal and human life, their energy usage over the past three millennia, culminating in the current expansion of human population and widespread application of easy-to-use fossil fuels but followed by a plethora of new human energy technologies.

It describes these technologies, their history ingenuity and solutions based upon engineering and science, recent usage, all designed to reduce reliance on fossil fuels and develop more environmentally viable, efficient energy usage.

The book concentrates on energy developments and social energisation particularly in Great Britain and Ireland, North America and China, in order to make comparisons. These comparisons limit the international range of the book but enable the author to keep it to a reasonable length. A much longer version would be needed to be exhaustive in international sources, but the author hopes that someone else may attempt that in the future.

I consider the stark consequences of carbon dioxide, CO_2, methane, CH_4 and other climate change emissions and their effect on Earth's future habitability. It shows some of the methods that are developing for cleaner and more sustainable futures, describing the present status of world energy usage.

I review changing energy demands through time, placing those developments in context alongside population growth and present usage by geographical area and source, including fossil fuels, hydroelectricity, renewable and nuclear energy developments, which have increased our technological awareness of the human impact on the environment.

In this respect I have been encouraged by Cardwell's '*History of Technology*' [12] which described the development of modern technologies from the time of the Greeks, through European science to a description of the Scientific and Industrial Revolutions, paying particular attention to Energy and Electricity. Some readers may not agree with my inclusion of nuclear power in a book on Clean Energy, but my view is that fission and possibly fusion could be important components, alongside renewables, for our energy future.

I present energy data by means of brief case studies, many diagrams and photographs, to show energy changes and the people that made them in different parts of the world, placing them in context. This was inspired by the work of the late Prof David McKay [21] in his exciting book '*Sustainable Energy - Without the Hot Air*', Vols. 1 and 2. I also add vignettes, from my own experience, that highlight individual improvements in clean energy usage.

The Earth's and Human timelines are provided in the Appendices together with a historical perspective, including pictures and diagrams.

I do not intend this book to be a climate-change polemic or to proselytise specific new energy methods but to demonstrate that human beings have extracted energy from their environment in a variety of ways over millennia, using knowledge to improve their methods and science, changing dramatically since the seventeenth–twentieth centuries scientific revolutions. However, over that time many new methods, especially the burning of fossil fuels, coal, oil and gas, have damaged the Earth's environment.

Scientific and engineering knowledge can improve our future environment which, with new technologies, could allow human societies to achieve a much lower carbon footprint.

Up to the beginning of the nineteenth century the development of these technologies was dominated by men, but from the nineteenth to twenty–first centuries this has changed significantly, women now play an increasingly important role in renewable developments, and I want to record that in this book.

Clean energy is a hugely complex issue, it is notoriously difficult to predict the future, particularly in this arena, where geo-politics, technology and markets can change so rapidly.

So, this is not intended as a book of predictions but to show that changes are happening and that the future can be influenced and guided by experience, scientific reason, good governance and imagination to resolve the impact of human energy usage on our planet.

An important aid in this matter has been Hans Rosling's helpful book '*Factfulness*', [29] which sets some of our World's data in context.

On a cautionary note, the Editors of Cardwell's '*History of Technology*' [12] decried books on history for their '*dreary recitation of facts*' and that is a problem for an author of Clean Energy,[4] as facts are essential for identifying future successful solutions, it is just their dreary recitation that I hope to avoid!

I intend to be neither alarmist nor to minimise the risks associated with current climate issues.

My practical experiences in the electrical and renewable industries lead me, in this book, to encourage a holistic and technically informed approach to the issues, raising confidence that climate-change issues can and will be resolved, with skill, wisdom and methods accrued from previous and current knowledge, including:

- Meteorology and Environment;
- Thermodynamics and Heat Management;
- Mechanics and Hydrodynamics;
- Electromagnetics, Semiconductors, Photovoltaics and Instrumentation;
- Networks and Computing;
- Connectivity and Social Responsibility.

Such approaches will need development and expansion for clean energy's successful future, while confirming they are sustainable.

However, there could be unforeseen disruptive influences on these changes and this book intends to show the diversity of issues around them, demonstrating how resilience should be improved.

Peter Tavner, Cambridge, 2023

Acknowledgements

I am indebted to several colleagues who have contributed to this book.

My friend, Dr Christopher Derrett, a former colleague at the MRC Air Pollution Unit, who encouraged me to take this task on.

Long-term colleague, Prof Bill Drury [20], for a perspective on historical technology set out in his comprehensive book '*The Control Techniques Drive and Controls Handbook*', 2nd Ed, 2009 [20]. I continue using his methodology in the Appendix to this book, placing engineering developments in context alongside Earth's history.

Another friend, Michael Carmichael, a former teaching colleague in Nigeria, for showing me his beautiful connectivity images, the book's frontispiece and endpiece, artistic representations of our Earth's interconnectivity and for introducing me to Prof Donald Cardwell's book '*History of Technology*' 1994 [12], which essential reading.

Special thanks again to an old friend, Doug Warne, who gave me sound research and editorial advice during my years at Durham and to the present IET Commissioning Editor, Christoph von Friedenburg, both of whom encouraged me in writing this and previous books.

I am also grateful to Dr John Moffett, Librarian of the Needham Research Institute in Cambridge for his help in locating detailed photographs, diagrams and energy developments in China during Prof Joseph Needham's technical work there [11] for inclusion in this book, as a memory in the streets of Cambridge of Prof Needham, when I was an undergraduate.

Finally, I acknowledge Prof Percy Hammond, my PhD Supervisor, whose inspiration and beautiful writing style taught me electromagnetism and history and Prof Kurt Schwarz, my former Technical Director, who taught me how to apply that knowledge and its consequent logic to electric machinery and drives, which will be essential for success in engineering renewable energy.

Connectivity frontispiece, courtesy Michael Carmichael

Nomenclature

Symbol	Explanation
A	Availability, $A=MTBF/(MTBF+MTTR)$
Acc	Acceleration factor for accelerated life testing
AEP	Annual energy production, MWh
C	Capacity factor, %
F or F^{-1}	Forward or backward Fast Fourier Transform
H	Efficiency
H_s	Wave height for sea-state
ICC	Initial capital cost, £
I	Drive train inertia, kg m^2
K	Constant in power balance equations
ku_n	Turbulence coefficient at a wind speed u of n, m/s
M	Instantaneous repair rate, $1/MTTR$
N	Speed of a machine rotor, rev/min
n	Number of years
P	Power, Watt
P_{det}	Probability of detection of a fault
p	Integer number of pole pairs in a motor or generator
Q	Heat flow, Watt/m^2
R	Resistance, Ohms
$R(t)$	Reliability or survivor function of a population of sub-assemblies as a function of time, failures/machine/year
r	Discount rate, %
S	Specific energy yield, MWh/m^2/year
Σ	Wind speed standard deviation
T	Torque, Nm
T	Temperature, °C
ΔT	Temperature rise, K or °C
T	Period of a wave, s
U	Wind speed, m/s
u_{ref}	Reference wind speed, m/s
1Λ	MTBF of an assembly or sub-assembly, h
V_{ref}	Mean wind speed at wind turbine hub height, m/s
V	RMS voltage, V
W	Work done in a wind turbine drive train, MWh
ω	Angular frequency, rad/s

Abbreviations

AC	Alternating current
BWR	Boiling water nuclear reactor
CANDU	Canadian Deuterium Moderated Nuclear Reactor
CAPEX	Capital expenditure costs
CCGT	Combined cycle gas turbine
CEFAS	Centre for Environment, Fisheries and Aquaculture Science
CEGB	Central Electricity Generating Board, former UK nationalised electricity utility
COP	Coefficient of performance
COSAG	Combined steam and gas turbine
CSP	Concentrated solar power
DC	Direct current
EHV	Extra high voltage, 345–765 kV
EMEC	European Marine Energy Centre
EOR	Enhanced oil recovery
EPICA	European Project for Ice Coring in Antarctica
EU	European Union
GRT	Gross Registered Tonnage
HAWT	Horizontal axis wind turbine
HM	Health monitoring
HV	High voltage, 100–345 kV
IAEA	International Atomic Energy Agency
ICF	Inertial confinement
IEC	International Electrotechnical Commission
IET	Institution of Engineering and Technology, formerly IEE
IEEE	Institute of Electrical and Electronic Engineers
IGBT	Insulated-Gate Bi-polar Transistor
IP	Intellectual property
JET	Joint European Torus
LCOE	Levelised cost of energy
LV	Low voltage, 100–1,000 V
MTBF	Mean time between failures, h
MV	Medium voltage, 1–100 kV
OREC	UK Offshore Renewable Energy Catapult, formerly National Renewable Energy Centre (NAREC)
PHES	Pumped hydro energy storage
PE	Photoelectric
PV	Photovoltaic

(Continues)

(*Continued*)

PWR	Pressurised water nuclear reactor
RAF	Royal Air Force
RPK	Revenue per passenger kilometre
SCADA	Supervisory control, or signal conditioning, and data acquisition
SPT	Solar power towers
TEFC	Totally Enclosed Fan-cooled Motor
USAAF	United States Army Air Force
VAWT	Vertical axis wind turbine
VVER	Water to water nuclear reactor

Chapter 1
Introduction

1.1 Scene setting

1.1.1 Background

This brief Introduction will set out the energy scene for this book, explaining the book's goal to increase our historic knowledge of human energy use, building up to the present state of energy conversion and setting out desirable future energy profiles.

The book will define 'clean energy', describing present significant Energy Vectors.

This will then lead to considering Earth's energy past, the changing effects of human energy usage with time and their impact on the world, its geographical and local environments.

The book will then describe how humans are attempting to modify their energy usage and their influences upon the environment and suggest the approaches humans could be taking to secure better human energy profiles for the Earth.

1.1.2 Definitions

Plants: all stationary living entities on the Earth's surface or within its rivers, seas and oceans.
Animals: all living creatures, mobile or otherwise, except humans.
Humans: Homo sapiens of all tribes, races, nationalities and affiliations.
Energy vectors: or energy sources, such as the Sun, coal, oil, gas fuels and electricity.

- **Sun:** electromagnetic solar power incident on the Earth and its capture by various biological and technological means.
- **Coal:** carbonised dead plant life compressed into the Earth's crust, a valuable but polluting Energy Vector.
- **Coal-gas:** a by-product of the conversion of coal into coke for the manufacture of steel, another valuable but polluting Energy Vector.
- **Gas:** vapourised carbon products associated with coal, but primarily from carbonised compressed reservoirs of dead marine animal life, another valuable but polluting Energy Vector.

- **Oil:** carbonised dead marine life compressed into reservoirs in the Earth's crust, another valuable but polluting Energy Vector.
- **Electricity:** an Energy Vector first developed by humans in the nineteenth century, defined as transmitting energy electromagnetically, via lines, cables, waveguides or the Earth's atmosphere, from power sources, whether battery or generator driven by an engine or turbine and supplied to industrial, commercial or individual users.
  - **Fossil-fired electricity:** Has been generated by engines powered by polluting Energy Vectors, coal, oil, gas or variants thereof.
  - **Nuclear-fired electricity**: Generated by turbines powered by nuclear fuel.
  - **Non-fossil electricity:** Now can be generated by non-polluting Energy Vectors, such as solar, wind and water (wave or tidal) power devices.
- **Clean and non-fossil energy sources:** defined as those sources which do not involve the introduction, directly or indirectly, of deleterious compounds into the atmosphere, soil or watercourses and do not leave behind residues of those compounds. The energy obtained from all these sources and any consequent pollution or lack of it.

1.1.3 Earth's history

Figure 1.1 shows a dramatic, graphic of the Earth's development up to the present time; it is developed in a form of 15 spiral loops, which emphasises the Earth's longevity.

The age in which humans are currently living is the Holocene Epoch at the end of the Quaternary Period, in the 15th and the uppermost loop of the spiral, seen at the lower part of the graphic, close to the caption, emphasising how short is humans' longevity compared to the Earth's longevity. The figure emphasises:

- The immense time-period which has elapsed since the Earth's formation, 4.5 billion years ago;
- The relatively short period in which the Earth has supported life, from the ninth loop of the diagram, 2 billion years ago;
- The arrival of the dinosaurs in the Jurassic period, in the 15th loop of the diagram, 251 million years ago;
- The extremely short period during which *Homo sapiens* has influenced the Earth's development, also at the most recent part of the 15th loop of the diagram, 300,000 years ago.

The following sections give a brief overview of plant, animal and human effects, giving rise to geographical, agricultural and industrial changes on Earth and starting to explore humans' energy usage.

1.1.4 Geology

The geology of our world is very important; it was formed in the earlier spirals of Figure 1.1, being effectively established by the Jurassic period.

Figure 1.1 Earth's development up to the Holocene Epoch, showing long preceding periods and the most recent human presence. Courtesy: J Graham, W Newman and J Stacy, US Geological Survey.

Professor Fortey [18] writes:

> Geology underlies everything, it founds the landscape, dictates the agriculture, determines the character of villages. Geology acts as a kind of collective unconscious for the World, a deep control beneath oceans and continents.

1.1.5 Plants

The first plant species to appear on Earth was more than 500 million years ago. There has been intense competition between plants. A very large variety of adaptable plant types have been able to evolve because of the following:

- Struggles between plant species to gain supremacy in specific environments, sometimes affecting human activity, for example in Europe the invasion of foreign species such as Hogweeds and Chinese Bush clover;
- Struggles between heath and forest as different plants compete for land resources;

- Struggles between growth near the sea, rivers and watercourses;
- The influences of animal vs human exploitation of plants and the development of human agriculture.

1.1.6 Animals and humans

The first animals appeared on Earth 700–800 million years ago.

Homo sapiens, humans, emerged around 300,000 years ago in Africa, evolving from Homo heidelbergensis. The spreading migration of Homo sapiens is shown in Figure 1.2(a), and the current human population distribution is shown in Figure 1.2(b), from 2005.

All animals, including humans, are continually processing food from their surroundings, to sustain life and growth. They are feeding from plants in the

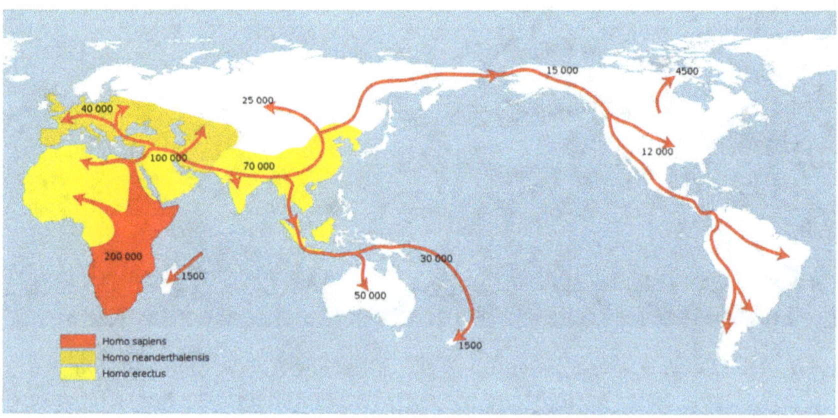

(a)

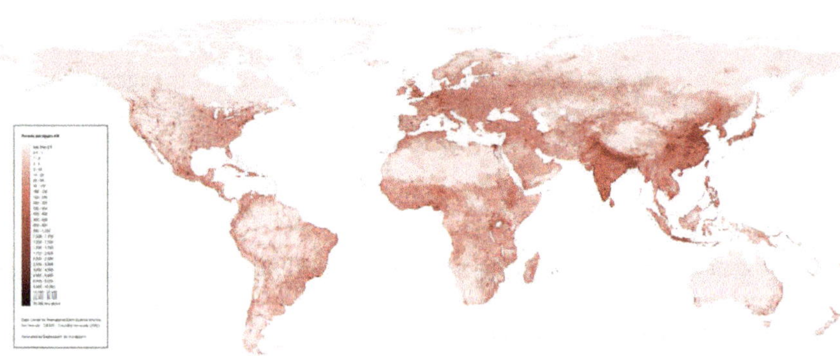

(b)

Figure 1.2 Human presence on Earth. (a) Migration map for humans, showing approximate years ago when migrations occurred; and (b) current human population density distribution map, 2005.

environment and off one another, using behaviours developed and evolved over millennia.

In the past this had generally been a generally sustainable process, constrained by limits to animal and human populations and the limitations of distance in and between the Earth's continents.

Earth's environment has been altered and developed by plant, animal and human processes and not necessarily always in a beneficial way.

Evidence from geographical history shows that there have always been historical limitations to the process of the spread of plants and animals; consider the following obvious examples.

Human colonisation has been limited by, for example, the following:

- In central and northern Russia, the forest and tundra have limited human occupation;
- In Greenland, an island part of the Danish Kingdom, located east of the Canadian Arctic Archipelago, between the Arctic and Atlantic Oceans and the world's largest island. It is three-quarters covered by the only permanent ice sheet outside Antarctica. Human colonisation in West Greenland ended during the late Viking age, as a result of a reversal of local warming, due to climate change. It now has a limited human population of 56,081. Nine-tenths of Greenlanders are of Inuit extraction, identifying themselves as Kalaallit (West Greenlanders), Inugguit (from Thule district) or Iit (East Greenlanders) depending upon their region.
- In northern parts of West Africa, due to southern expansion of the Sahara Desert;
- In Central Africa dense forest limits the penetration of human habitation, although several Pygmy tribes survive in small ethnic groups in Rwanda, Burundi, the Congo, Central African Republic, Cameroon and Equatorial Guinea;
- In north-west China, north of Lanzhou, Gansu province, due to the southerly expansion of the Gobi Desert;
- In Australia, the continent's Great Red Centre gives limited support to human life. But transient Aboriginal communities, with intimate knowledge of local flora and fauna and experience of the seasonal weather conditions, can survive and migrate through these areas;
- In Antarctica, the Earth's most southerly and unpopulated continent and the last great wilderness on Earth.

Predation within humanity itself, for example the following relatively recent examples:

- Human movement southwards through the Americas from 15,000 years ago;
- Then, from the fifteenth century onwards, the subsequent colonisation from Europe of the Americas;
- In the seventeenth century, the colonisation of Southern Africa from Europe, then consequential wars between the indigenous African and incoming

European peoples and subsequent introduction of labour colonisation from the Indian subcontinent;

- The influence of European colonisation in Australia and New Zealand from the eighteenth century on the pre-existing indigenous Aboriginal and Maori peoples, respectively.
- The wars in Papua New Guinea between indigenous human tribes during the nineteenth and twentieth centuries;
- The two world wars of the twentieth century, driven largely by local European expansionism, exciting similar action in the Far East by Japanese occupations of China and South East Asia;
- Colonisation by invasive animal species, for example the grey squirrel in Europe or by the European rabbit in the Southern hemisphere in Australia, as a consequence of human immigration;
- Invasive effects of human society in general on the balance between forest, heath and pasture in various countries, as humans cut down trees for timber and the extension of farming land, for example the loss of rain forest in Brazil and in Central Africa.

1.2 Energy and climate change, a short history

The development of agriculture, industry, engineering and science between the eleventh and fifteenth centuries exemplifies an expansion of human influence on our Earth's environment.

Cardwell [12] wrote that at the time of the Domesday Book in the eleventh century (1084), England, had up to 5,000 established watermills providing milling power for a population of about 3 million. Therefore, he said the 'gross horsepower per capita' was relatively high at that time, which stimulated technical advances in both farming and manufacturing.

However, in the fourteenth century that pace slackened, for reasons that are unclear to historians, but possibly associated with a series of plagues, including the Black Death of 1348 affecting the whole of Europe.

By the fifteenth century, the pace of technical advances in science and engineering throughout Europe had started to pick up again.

In respect of climate change in 2023, Michael Wood, a British historian, has given a powerful historical insight into our appreciation of the effects of energy change.

He described how in the eighteenth century the English naturalist and clergyman, Gilbert White, recorded in his book '*The Natural History of Selbourne*', published in 1789 [1], the effects on his local UK Hampshire environment of what local farmers called 'Blue or London Smoke', a stinking airborne pollution, caused by coal burning in the capital.

Wood then linked this observation to a historical global event in the eighteenth century, the 'Great Divergence' in which the Industrial Revolution, fed by coal burning, enabled the British economy to overtake that of China.

This was exemplified during the United Kingdom's Mission by Earl George McCartney in 1792–94 to meet the Qing Dynasty's Qianlong Emperor in Beijing (北京).

Also, in the nineteenth century the Irish scientist John Tyndall, working from his knowledge of glaciology and investigating the absorption spectrometry of gases, began in 1859 to make the link between atmospheric CO_2, due to coal burning, and what we now call the 'Greenhouse Effect'.

At the beginning of the twentieth century, 1912, an article appeared in the magazine 'Popular Mechanics' saying:

> The furnaces of the world are now burning about 2,000,000,000 tons of coal a year. When this is burned, uniting with Oxygen, O2, it adds about 7,000,000,000 tons of Carbon Dioxide, CO2, to the atmosphere yearly. This tends to make the air a more effective blanket for the earth and to raise its temperature. The effect maybe considerable in a few centuries.

Then in 1958 American scientist Charles Keeling began studying air quality at a Hawaii observatory, measuring CO_2 accumulations in the atmosphere. This developed into the Keeling Curve, which gives a visual record of world atmospheric CO_2; see Figure 1.3(a).

Sprague de Camp [9] has given another version, from a US viewpoint, of the effects of energy changes.

By the 1980s the science was clear and NASA Chief Scientist, James Hansen, in 1988, during a Congressional testimony, famously broke ranks on climate change to warn a US Senate Committee of impending catastrophe; see Philander [14].

> Michael Wood's analysis has led to this insight on climate change:
> What we did before in ignorance, we now do in full knowledge.

Figure 1.3(b) shows the changes that have happened in the United Kingdom reducing our CO_2 emissions. Some of this is due to the decline in our coal industry and the offshoring of manufacturing but it is also due to efforts that have been made to change from fossil fuel-fired to renewable electricity production.

For an alternative historical view of climate change, see Frankopan [37].

A graphic of recent climate change events is shown in Figure 1.4, which includes wild weather and the Japanese tsunami that damaged Fukushima's nuclear power stations. Strictly some of these incidents are unpredictable geophysical events, which may not be initiated by human energy exploitation, with which this book is concerned.

However, the graph does demonstrate that once human beings focus on a problem, we can gather valuable information, the difficulty is interpreting that data correctly and acting upon it.

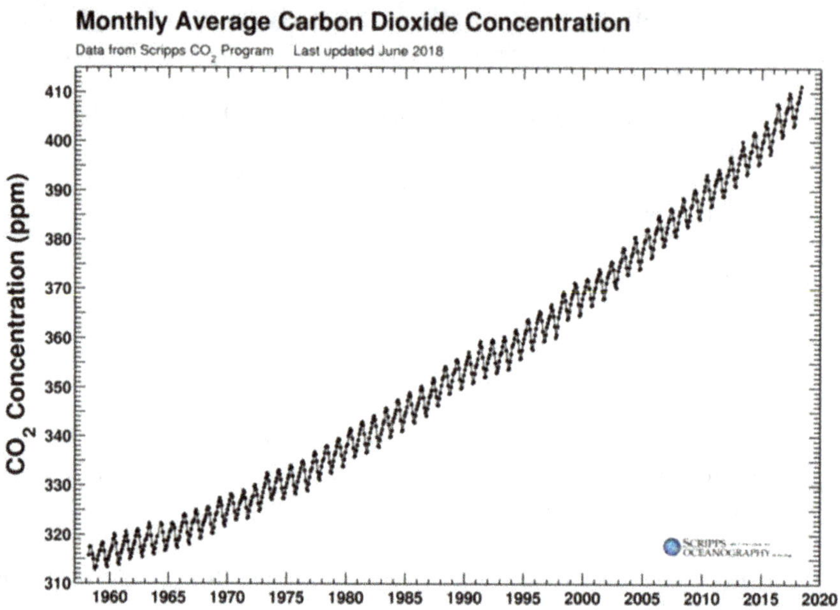

(a)

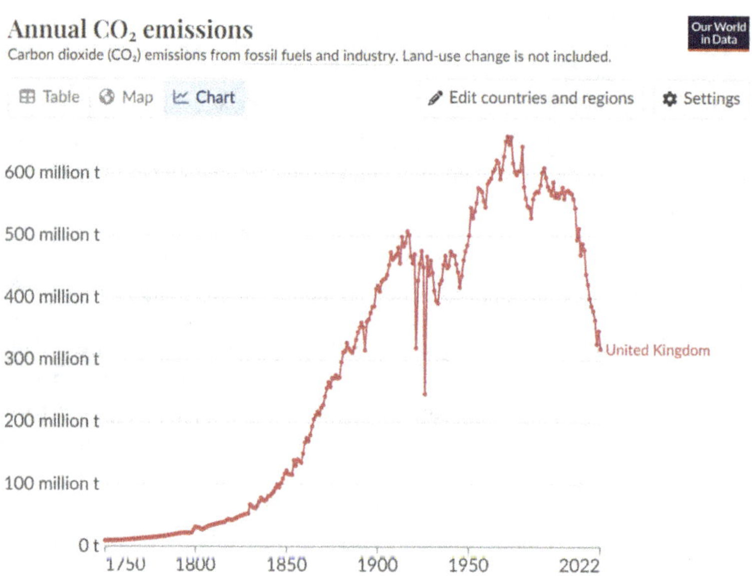

(b)

Figure 1.3 Global Warming and atmospheric CO_2. (a) Earth's atmospheric CO_2 content. Twentieth-century Keeling Curve, 1965–2022, Courtesy: Scripps Institution of Oceanography, University of California, San Diego; (b) United Kingdom's CO_2 emissions, 1750–2020, Courtesy: Max Roser, Our World in Data.

1.3 Current human energy usage

1.3.1 By fuel and geographical area

The historic changes in world energy use over the last 221 years are shown very clearly in Figure 1.4, divided into colours according to fuel source.

The following issues arise from the data in Figure 1.5:

- The importance of traditional biomass energy consumption, which remained sensibly constant from 1800 to 2021 between 5,000 and 10,000 TWh;
- This biomass consumption is at about the same level as coal energy consumption in the West from 1890 to 1930;
- From 1850 onwards, there was a clear and steady rise in coal consumption;
- However, that is now starting to fall;
- From 1925, there was a steady increase in oil usage;
- Then from 1950 a similar steady increase in natural gas usage;
- From 1950, a small but steadily increasing use of hydropower;
- From 1980, the beginnings of influence from other renewable sources, wind, solar and modern biofuels, but again rather small in quantity.

This graphic, Figure 1.5 puts the world's energy challenge firmly in perspective, the following sections help to explain how these increases have come about as development has increased in Asia and industrial production has moved from West to East.

Figure 1.6 gives a regional insight into current fuel usage, as follows:

- Asia Pacific usage is rising rapidly above 70,000 TWh per annum, primarily because of industrial growth in India, China, Japan and South East Asia;

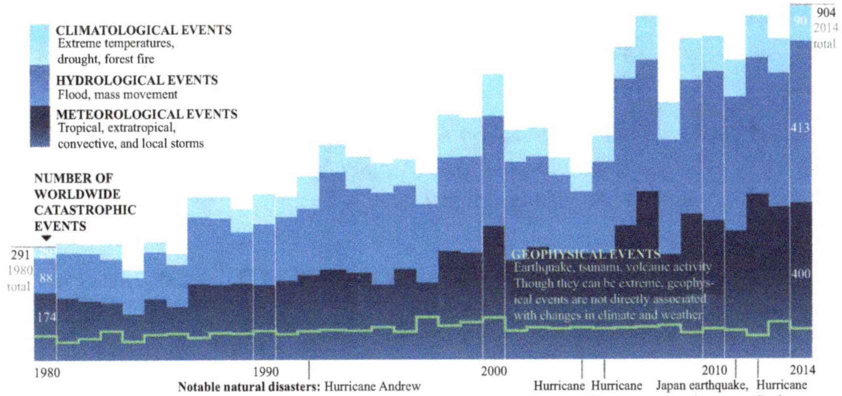

Figure 1.4 Climatological, hydrological, meteorological and geophysical incidents influencing climate change, 1960–2014. Courtesy: National Geographic.

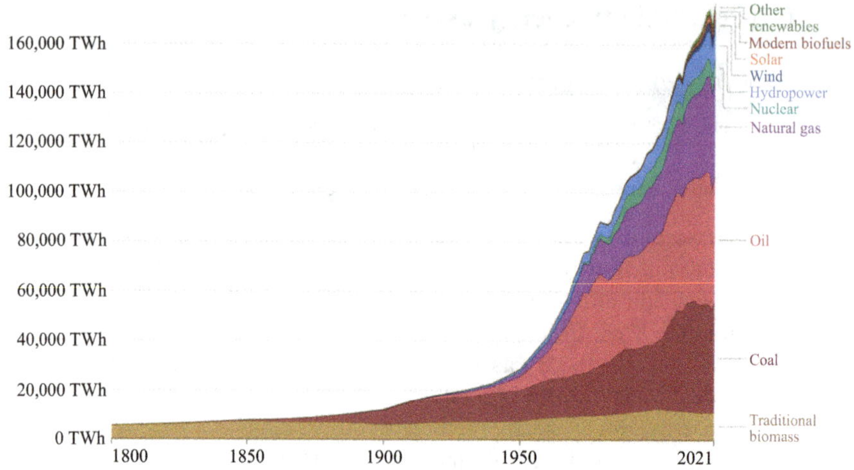

Figure 1.5 Changes in global energy consumption from different fuels, 1800–2021. Courtesy: Vaclav Smil and BP Statistical Review of World Energy (2021).

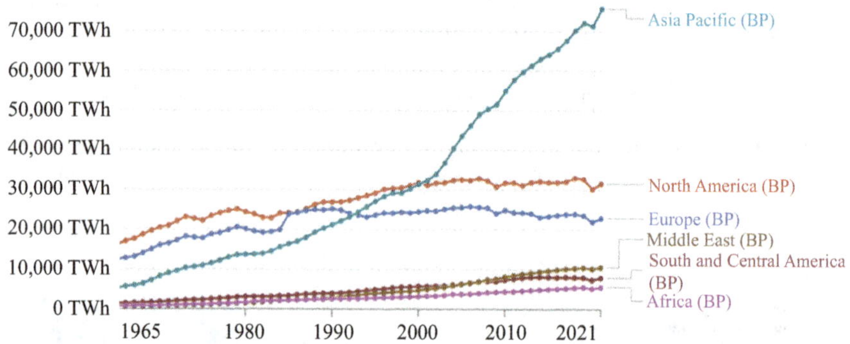

Figure 1.6 Recent changes in use by fuel and geography, 1965–2021; global energy consumptions by geographical area. Courtesy: BP Statistical Review of World Energy (2021).

- North America's and Europe's usages, between 15,000 and 30,000 TWh per annum, have started to decline slowly.
- This is because considerable industrial activity in those areas has been passed onto the Asia Pacific region;
- The Middle East, South and Central America and Africa, below 10,000 TWh per annum, currently remain at a relatively low level;
- It is notable that Africa has an energy usage of less than 5,000 TWh per annum, despite its large population, reflecting less industrially oriented economies. Perhaps this will be an area of significant change in the future.

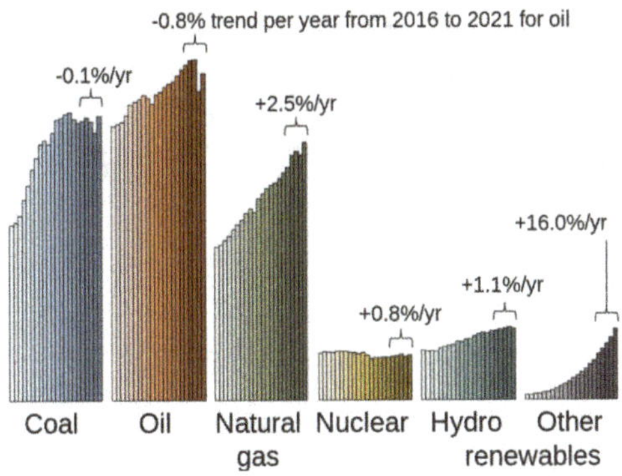

Figure 1.7 Global energy consumptions of different fuels or Energy Vectors, 2000–21. Courtesy: Friedlingstein (2021).

1.3.2 By usage and energy vectors

Figure 1.7 gives a recent, more up-to-date and sharper insight into fuel usage by Energy Vector, as follows:

- Coal usage is starting to decline;
- Oil and natural gas usage continues to grow;
- Nuclear power usage remains relatively constant but clearly has recent issues, influenced by world reactions to the nuclear accidents at Fukushima, 2011, earlier nuclear accidents at Chernobyl, 1986, and Three Mile Island, 1979;
- Hydropower usage is gradually increasing, but there is an environmental limit to the acceptable exploitation of riverine resources;
- The following renewable sources are steadily increasing in significance, as follows:
  - Wind;
  - Solar;
  - Modern biofuels.

But their contribution is still small compared to coal, oil and natural gas.

It has been suggested by Max Roser that the human world currently faces two energy problems:

- Most human energy production still produces greenhouse gas emissions;
- Whilst hundreds of millions of humans entirely lack access to that energy.

1.4 Possible future trends

1.4.1 Setting the scene

Experience of the last 100 years has demonstrated that human expansion of numbers and growth in economic development are having a serious influence on the

Earth's environment and on its other inhabitants, animals and plants. That is being caused by degradation of the physical environment atmospheric pollution, rising temperatures and rising sea levels.

Individual causes are complex, but these effects are primarily due to three factors:

- The number of humans on the planet;
- The exploitation and burning of fossil fuels;
- Progressive evolution of the plants, animals and humans of our planet.

Ancient historic information about temperatures and sea levels suggests that human intervention is not the only driver.

However, the human causes mentioned earlier are being understood and can be modified in the future, based upon scientific understanding.

1.4.2 Food

Humanity has found ways to feed itself on Earth, human hunger has fallen to much lower levels than in previous centuries, largely due to:

- Improved effectiveness of mass agricultural due to:
  - Greater land exploitation;
  - Use of more effective fertilisers;
  - Access to more efficient transport links that carry food from production areas to demand locations, by land sea and air.

There have been consequences from these successes:

- Much more intensive farming, which can be damaging to the environment causing consequential loss:
  - Of other Earth plant species;
  - Of other Earth animal species.
- Consequent more intensive use of fossil fuels for:
  - Land exploitation;
  - Intensive fertiliser application;
  - Widespread use of fossil fuels for transportation by land, sea and air.

However, each of these successes has deleterious effects on the environment, albeit modifiable with understanding, the most important of which is to change from intensive use of fossil fuels towards more environmentally acceptable processes.

None of this is easy and requires time to reorganise and rebalance.

Humanity has already started making changes; we just need to make better and more coordinated progress.

1.4.3 Light

Lighting technology changed in the second century to the burning of rush oil lamps, then in the twelfth century to using candles, in the eighteenth century to burning oil, then in the mid-nineteenth century to burning coal-gas. All these techniques are potential sources of pollution.

Since the eighteenth-century lighting has been an area of human technology application that has made enormous and life-enhancing progress, in terms of facilitating work, medical care and family life.

At the end of the beginning of twentieth century, we transitioned from almost all those outdated methods to using electricity for lighting, generated by relatively low-efficiency coal-fired power stations. These electrical systems energised low-efficiency light bulbs, progressively providing factories, schools, homes, city roads, rail and road transport with light.

The average efficiency of modern electrical lighting methods at the start of the twenty-first century has gradually improved some orders of magnitude better than the low-efficiency light bulbs used 100 years ago.

However, the easy availability and cheapness of these new low-energy lighting methods have also encouraged a wider use and we still need to reach a balance, reducing illumination to what is really needed.

This is now gradually happening and is driven by the cost of lighting.

1.4.4 Transport and power

Much of our transportation still depends upon hydrocarbon fuels, in particular road transport and aviation applications.

However, considerable inroads into those fuels, largely the result of the need for higher efficiency, have been made in the last 100 years, the following are examples:

- The expansion of national and international electric railways in Europe, India, Russia, China and Japan;
- The development of electric vehicles, so far mainly for family cars but gradually expanding into larger commercial applications such as vans, trucks and heavy-duty lorries;
- Very significant improvements in the performance and efficiency have been made in:
  - Electrically powered national rail and city transportation railway systems;
  - Petrol-driven cars and vans;
  - Hybrid-driven, petrol and electric, cars and vans
  - More efficient diesel trucks and railway locomotives;

Aviation has experienced since the First and Second World Wars very considerable expansion but with an impact on pollution. Longer distance air travel has opened international markets but is reliant on polluting, high-energy aviation fuels. However, the net effect of aviation on CO_2 emissions has remained relatively low at 2%–3%. But we have become more reliant on faster services, using more powerful polluting jet engines.

This culminated in the Concorde super-sonic passenger jet aircraft with its prodigious carbon footprint, which proved to be valuable for very high-speed transits but was not economically viable for widespread aviation.

Modern aviation fuels have moved gradually from high octane aviation fuel, Avgas, towards lower cost jet fuels, Jet A, A-1 and Jet B, more akin to paraffin.

Work is now progressing to develop less fuel-intensive engines, incorporating traditional propulsors with the use of electrical power. But that may mean a move towards lower aviation speeds to facilitate the transition.

Power generation in the nineteenth century relied heavily on coal with some wood or peat burning. This was very polluting and affected large city atmospheric conditions, first in Western Europe, then in South Asia, China and then South East Asia.

Electric power generation expansion, by GWh, accelerated with the development of larger fossil-fuelled power stations in the early part of the twentieth century.

Large-area national electrical grids and their interconnection then led, in the twentieth and twenty-first centuries, to the reduction of fossil fuel use by accessing extensively distributed more sources, including more efficient fossil and hydropower sources.

The ability to schedule electrical power distribution over very large areas, for example in Russia, the United States and China, has facilitated some reduction of polluting trends by spreading generating fuel sources and paying more attention to higher efficiencies.

1.4.5 Heat

A significant proportion of humans now live in large and distributed communities in both hot and cool environments, respectively, requiring cooling or heating assistance and a wide range of transport facilities, both until recently largely dependent upon fossil fuels in those infrastructures.

For example, 1.7 billion people, or 23% of Earth's population, live in cities with at least 1 million inhabitants. The following three points are relevant:

- Human technology is now such that the thermodynamic efficiencies of fossil usage could rise above 60%, reducing the effects of burning those fuels and therefore their amounts of pollution, especially if we could rapidly deploy those higher efficiency plants.
- However, there is also a wide range of renewable energy sources being developed that could allow us to move away from the use of fossil fuels for electricity and heating.
- In fact, we could effectively cease to place any reliance thermally on any fossil fuels, using them more frugally as chemical feedstocks, avoiding their contribution to atmospheric CO_2.

1.5 Summary

This chapter has demonstrated the decisive evidence of human impact on the Earth's climate, particularly over the last 100 years.

It has also shown that there is a recognition, at least in some human societies, of the potential consequences of climate change, in terms of damaged environments, more aggressive weather conditions, loss of polar ice caps and damage to human and animal habitats.

The initial human response has been vocal, particularly in more developed countries in Europe, North America and the Far East. However, poorer countries, such as Bangladesh, Vietnam and remote Pacific Islands, have also been vocal as they experience serious problems due to sea level rise and unexpectedly fierce storm conditions, such that lives and livelihoods have been put at risk.

Many declarations of intent have been made, but not so many intents have been brought into practice.

More could be done to help developing economies, where capital is not so readily available, to mitigate the effects, but who are naturally more concerned about growing their economies to match the larger economies.

Climate change has therefore become an important human political and economic issue worldwide.

As climate change knowledge increases, rational solutions are starting to emerge. Significantly a great deal of engineering work has been done to alter our methods of exploiting energy.

The current difficulties are that large fossil energy sources, like coal, oil and gas, have functioning, profitable extraction methods which release a great deal of capital to invest in development. For developing economies in poorer countries, these proven methods are a short-cut to industrial growth and increased national income.

However, developed countries have made considerable progress in technological areas to divert us away from fossil fuels, particularly in solar, wind and marine renewables.

The profitability of these technologies has steadily improved as they are proven in service, capital costs reduce, operational experience grows and running costs fall.

It is clear that this is a work in progress but that renewable energy sources can and will steadily improve, however.

This book will try to explain how that can be done, remembering that the profitability of renewable industry has so far fallen well short of the oil and gas industry.

Chapter 2

Ancient past, last 542,000,000 years

2.1 Introduction

This chapter sets the scene for current climate change issues from the point of view of the ancient past.

Figure 1.1 showed diagrammatically the development of the Earth's structure and associated geological, volcanic, climatic, flora and fauna changes it has experienced up to the historic entry of humans.

The intent of this chapter is to draw out the geological, volcanic, geographical and species population basis for later developments, showing the early effects on the Earth up to *Homo sapiens*' appearance.

Then it shows the effect of early human interventions that lead to our current concerns about climate change.

2.2 Earth's age and structure

2.2.1 Continental drift

Continental drift describes the way geologists believe that the Earth's continents have moved over time.

This historical development of the Earth's current continents is described through plate tectonics.

Figures 2.1–2.4 show graphically, in seven images, our current understanding of the Earth's present continent's formations and their movements over 600 million years.

The continents and their movements to their current locations are the key to understanding the location of valuable mineral resources within the Earth's crust.

The exploitation of these resources by man has had a significant impact on the Earth's current climate.

However, a deeper understanding of these resources, their current location, original location and subsequent movement may influence humans' ability to reduce and stop global warming.

This may be achieved by changing these mineral resource exploitations by more effective and less destructive interventions.

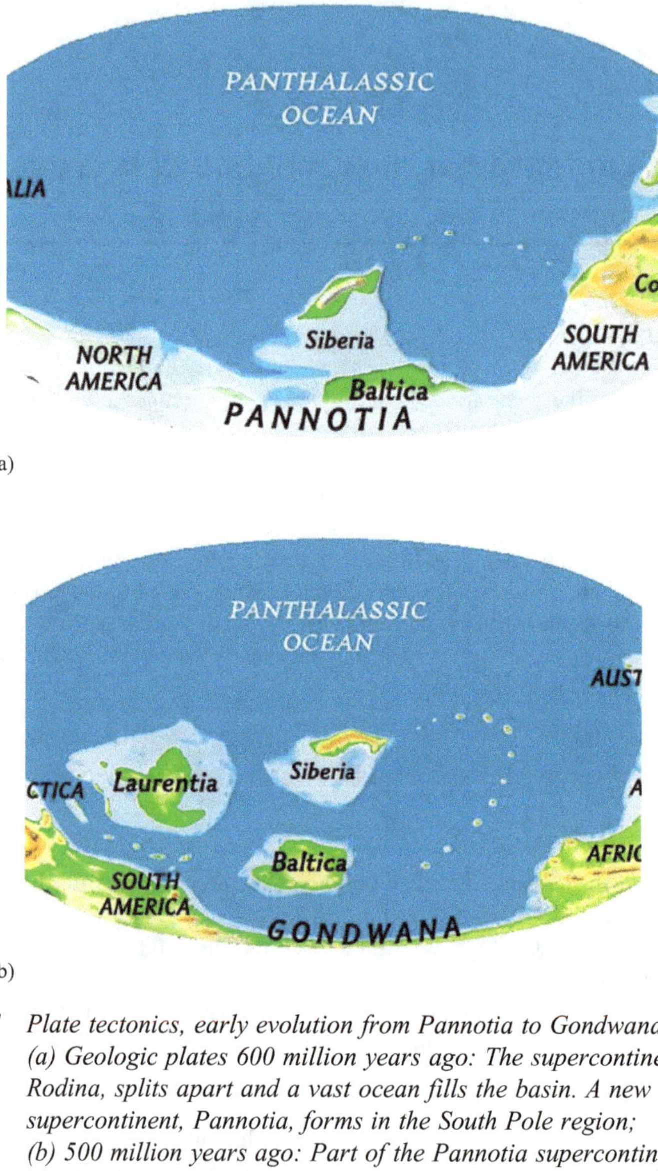

(a)

(b)

Figure 2.1 Plate tectonics, early evolution from Pannotia to Gondwana. (a) Geologic plates 600 million years ago: The supercontinent, Rodina, splits apart and a vast ocean fills the basin. A new supercontinent, Pannotia, forms in the South Pole region; (b) 500 million years ago: Part of the Pannotia supercontinent drifts North splitting into three: Laurentia, present-day North America, Baltica, present-day northern Europe and Siberia.

Figure 2.5 shows the original location of today's named continents in the ancient Gondwana supercontinent, showing the connected distribution of ancient animal and plant fossil remains across it from those ancient times, which correspond with their current locations in our present-day continents.

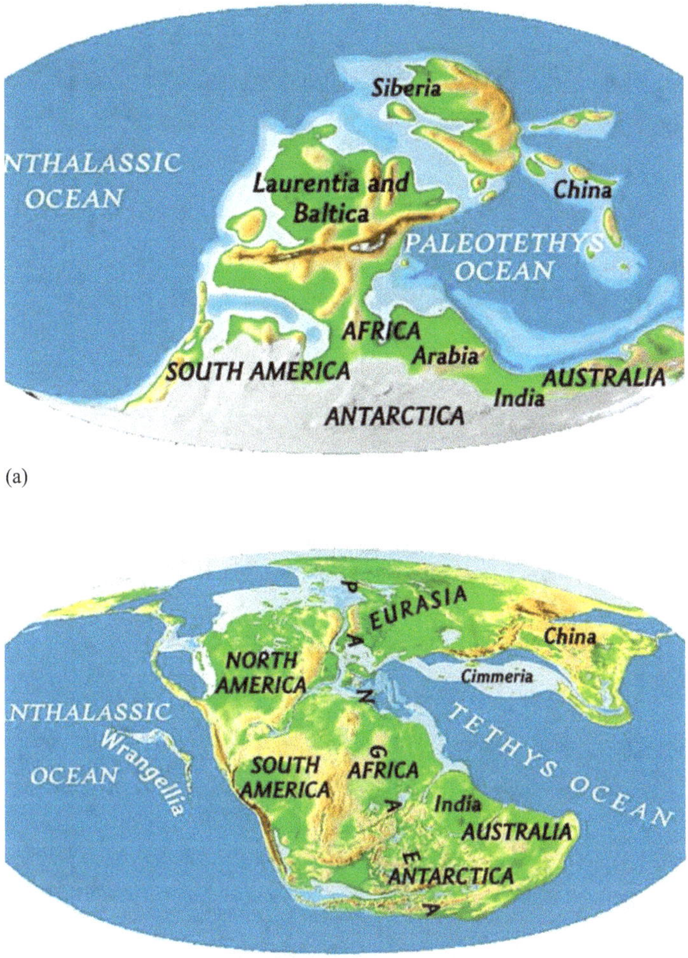

(a)

(b)

Figure 2.2 Plate tectonics, evolution from Laurentia and Baltica to Pangea to early present day, extinction events 1 and 2, see Figure 2.6. (a) 300 million years age: The Laurentia and Baltica landmasses collide; (b) 200 million years ago: The supercontinent Pangea forms. Courtesy: National Geographic.

This explains the presence of homogeneous fossils, for example Cynognathus or Mesosaurus or Lystrosaurus, our now widely dispersed across present continents.

Figures 2.1–2.4 spanned five main species extinction events, which are identified in the following timeline, Figure 2.6 as points 1, 2, 3, 4 and 5.

All these extinction events preceded current human occupation of our modern world.

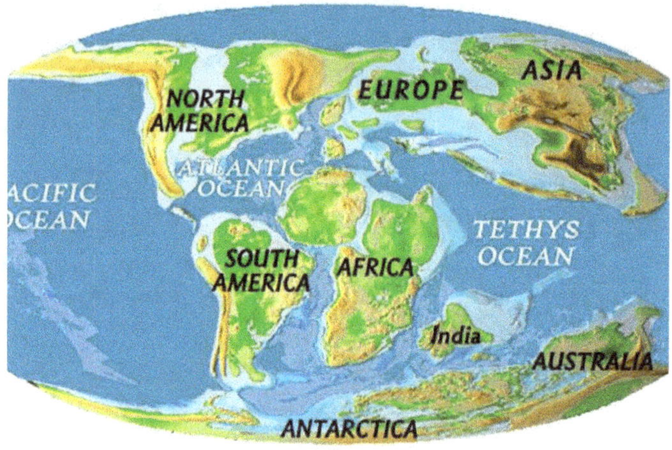

(a)

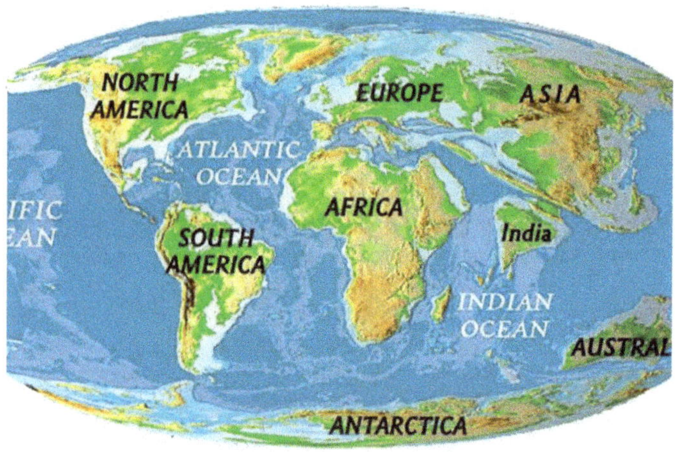

(b)

Figure 2.3 Plate tectonics, evolution from Pangea to early present day, extinction events 3, 4 and 5, see Figure 2.6. (a) 100 million years ago: In Pangaea violent eruptions extinguish life and break it apart, the Atlantic Ocean pours between Africa and Americas, India breaks away from Africa. Antarctica and Australia, connected above sea level, are stranded near South Pole. (b) 50 million years ago: Continental fragments collide, pushing up mountain ranges. Africa and Europe's collision raises the Alps, India and Asia's collision forms the Himalayas. Courtesy: National Geographic.

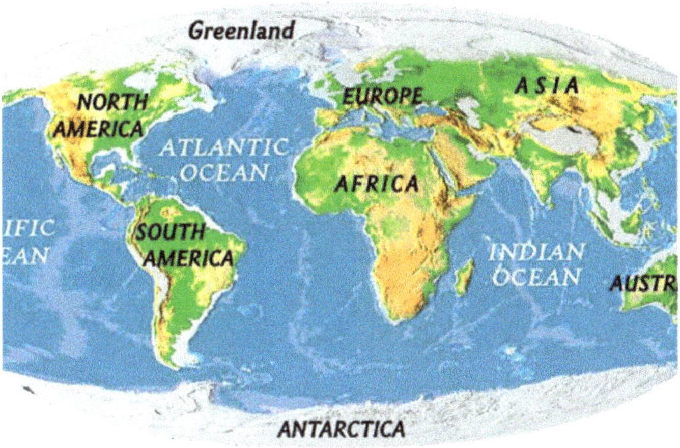

Figure 2.4 *Plate tectonics, the present day. Present day. Isthmus forms between North and South America, Australia and Antarctica split changing global ocean currents and climate. N American Great Lakes carved out by ice sheets, 20,000 years ago. Warmer temperatures melt ice, raising sea levels. Courtesy: National Geographic.*

However, in this book, we are concerned with the present day and the immediate future, considering the effects that our industrial and technological revolutions are having on our climate.

2.2.2 Recent geographical changes

In recent times, the changes to the Earth's environment witnessed by humans have been far less spectacular than the gigantic effects of plate tectonics, seen by preceding creatures, exemplified by the changes in Figure 2.5.

However, in our present world, continents continue to move, mountain ranges continue to rise, deserts continue to expand, for example:

- In Europe in the Alps, as Italy collides with the continent and more spectacularly in the Himalayas, as India crashes into Asia;
- In West Africa, ancient plains are steadily eroded by annual tropical downpours, whilst the Sahara Desert slowly expands southwards;
- In East Africa, the Great Rift Valley, a series of contiguous geographic trenches, extends from south to north 5,950 km from Lebanon through the Red Sea into East Africa down to Mozambique. Several deep, elongated ribbon lakes exist on the valley floor; the largest examples are as follows:
    - Lake Malawi;
    - Lake Rudolf;
    - Lake Tanganyika.

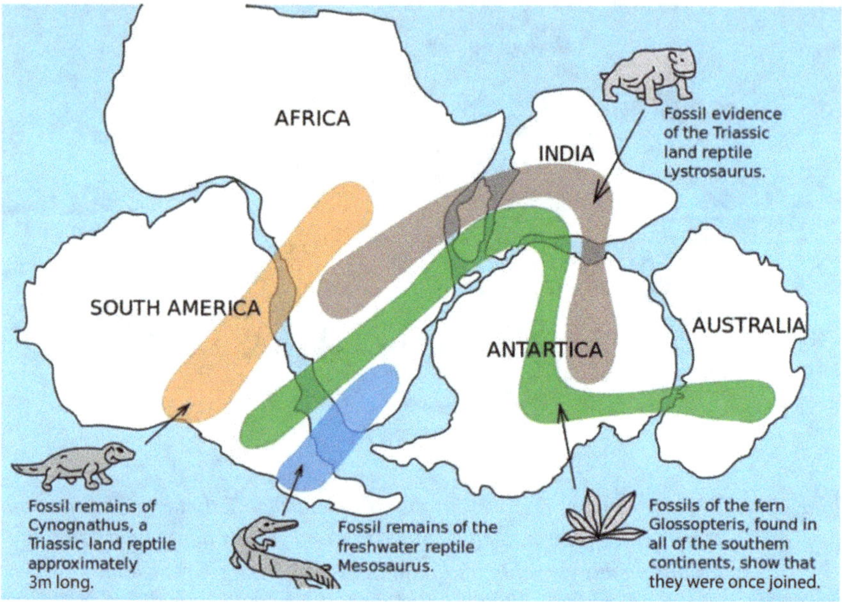

Figure 2.5 Ancient supercontinent Gondwanan map, migrating outwards to become the current-named continents, showing locations of related fossil remains. Courtesy: US Geological Survey.

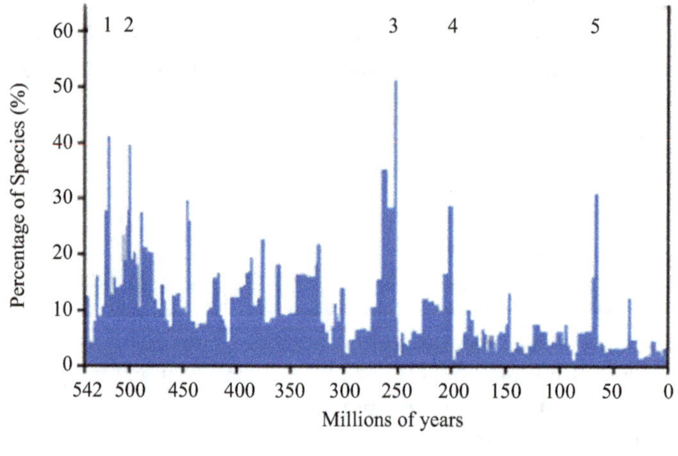

Figure 2.6 Earth's previous five main extinction events, over 542 million years.

- In West China, the Taklamakan Desert continues increasing in size, being surrounded by the Tien Shan mountains to the North and Kunlun Shan mountains to the South, the whole under pressure from India's motion Northwards into Asia.
- In Northeast China, huge ancient and static plains were being eroded by the wind; some rain and the exit across them of the Yellow River (黄河) to the North and Yangtze River (長江) to the South brought debris pouring out of the Himalayas.

2.3 Earth's atmosphere

2.3.1 Extinction events

Species become extinct as organisms compete for environmental niches, with some species succeeding and some fading, as genetic mutations lead new species to succeed over older ones or as environments change.

At long, irregular intervals the Earth's biosphere has suffered catastrophic die-offs, known as mass extinctions, comprising an accumulation of smaller extinction events and a mass extinction can occur over a relatively short period of time.

Figure 2.6 shows the apparent percentage of marine animal species becoming extinct on Earth, over a 542-million-year time frame. Only a few species are included in this diagram, that is, only those readily fossilised. The five largest extinction events are numbered in Figure 2.6, the first two Cambrian-age extinction events, on the far left, are large in percentile magnitude but small in absolute numbers, due to the low levels of fossil-producing life at that time.

The Earth is currently entering another extinction event, known as the Holocene; see Kolbert [26]. This has been triggered by much larger numbers of plant and animal species, including humans, than were previously experienced on Earth, all competing for the Earth's resources.

Human development, population growth and energy usage have therefore become a significant influence on the Holocene event's progress.

2.3.2 Temperature change timeline

A record of Earth's temperature variations over the last 800,000 years can be derived from ice cores taken by the European Project for Ice Coring in Antarctica (EPICA), Figure 2.7(a).

This shows a rising temperature difference, over a period of 800,000 years, and an uptick at the current time on the far right of the diagram, representing current global warming.

Figure 2.7(b) shows the Earth's temperature changes over the last 50 years.

A sixth large extinction event, now called the Holocene Extinction, is currently progressing beyond those extinction events shown in Figure 2.6, driven at least in the recent three centuries by human activity.

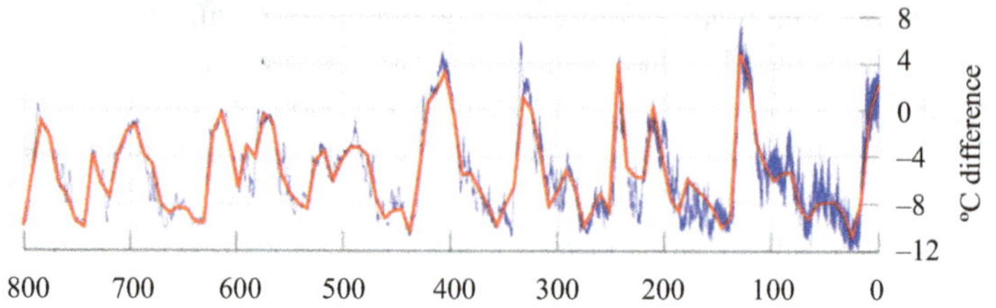

(a)

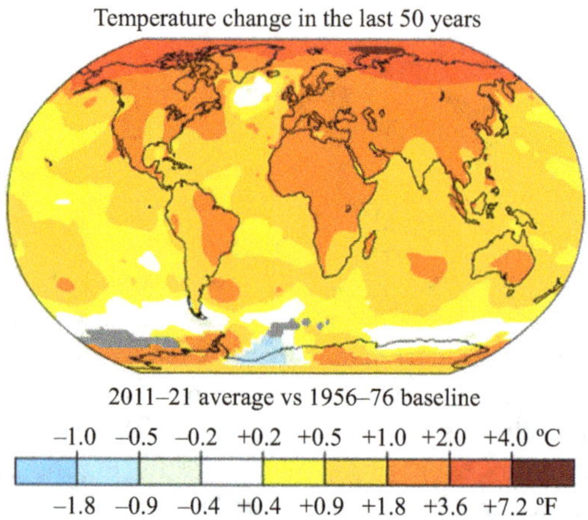

(b)

Figure 2.7 Earth's surface temperature variations. (a) Last 800,000 years Earth's surface temperature change. Courtesy: European Project for Ice Coring in Antarctica EPICA. (b) Earth's surface temperature changes 2011–21 versus 1956–76 baseline.

2.4 Human species

2.4.1 Human evolution

A diagram of the evolution of the human species is as Figure 2.8.

- *H. sapiens*, Latin for '*the wise man*', is the species to which all modern human beings belong. One of several species grouped into the genus Homo but not yet extinct. *H. sapiens* was applied, in 1758, by the founder of modern biological classification, Carolus Linnaeus.

  It had long been known that human beings physically resemble primates more closely than any other known living organisms, but at the time it was a daring act to classify human beings within the same framework used for the rest of nature. Linnaeus, concerned himself exclusively with similarities in

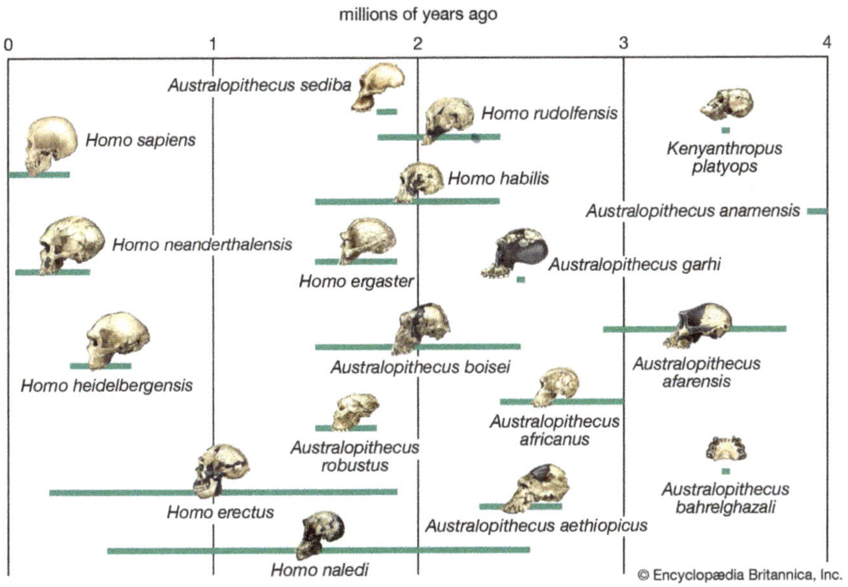

Figure 2.8 From right to left, evolution to Homo sapiens.

bodily structure, facing only the problem of distinguishing *H. sapiens* from apes, differing from humans in bodily as well as some cognitive features.

- *Homo neanderthalensis*, an extinct species of archaic human, genus Homo, known from fossils in Africa, Europe and possibly Asia, dating from 600,000 to 200,000 years ago.

  The name first appeared in print in 1908 to accommodate an ancient human jaw discovered in 1907 near the town of Mauer, 16 km southeast of Heidelberg, Germany. Among the fossils found with the Heidelberg jaw were those of several extinct mammals living about 500,000 years ago.
- *Homo erectus*, most likely originated in Africa, although Eurasia cannot be ruled out. An extinct species of the human genus, perhaps an ancestor of *H. sapiens*. The species seems to have dispersed quickly, starting about 1.9 million years ago near the middle of the Pleistocene Epoch, moving through the African tropics, Europe, South and Southeast Asia.
- *Homo naledi*, Latin and Sesotho mix for '*star man*', from Southern Africa. An extinct human species, initially thought to have evolved, from 2.8 to 2.5 million years ago, during the Pliocene and Pleistocene epochs, about the same time as genus Homo's emergence, their classification with other Homo species remains unclear. Subsequent fossil analyses, in the only known location where the remains have been found, suggest that *H. naledi* may be significantly more recent or have persisted deep into the Pleistocene period before dying out. In an excavation of the Rising Star cave complex in South Africa's Transvaal,

discovered in 2015, more than 1,500 fossil specimens were found. They have exhibited crafted stone tools, sophisticated burial arrangements and distinctive linear rock inscriptions at the burial locations.

2.4.2 Human interventions

2.4.2.1 Human prehistory

The origin of *H. sapiens* is believed to have been at the centre of the African continent more than 2 million years ago, most likely in the Kenyan Rift Valley, where the oldest known stone tools have been found.

There is some evidence of early movement from the Rift Valley towards the eastern coast of Africa, where communities would have been able to expand through the dual benefits of hunting game on-shore and catching fish along the coastline.

Early expansion out of Africa by *H. erectus*, by migration northwards, along the eastern coastal shoreline, thereby retaining land or seafood benefits, possibly occurred 2 million years ago.

We cannot know the population size of these migrations, which may have been as small as hundreds of people.

Nearly 1.9 million years ago, *H. sapiens* spread through East to Southern Africa, *Homo ergaster*, but not at that time into West Africa.

H. naledi appears in Southern Africa and is an extinct hominin species discovered in 2013 in the Rising Star Cave, so-called Cradle of Humankind, South Africa dating to the Middle Pleistocene 335,000–236,000 years ago. The initial discoveries comprised 1,550 bone specimens, representing 737 different skeletal elements and at least 15 different individuals. Despite this exceptionally high number of specimens, their classification with other Homo species remains unclear.

H. erectus migrated out of Africa via the Levantine corridor and Horn of Africa to Eurasia, believed to be 1.8 million years ago, being related to the operation of the so-called Saharan Pump, which pushed mankind out of the increasingly arid environment in Central Africa.

There were probably successive migration waves with some later migrations back into the African continent, based upon the recent location of significant archaeological sites with ancient human remains, see the list next.

Key sites identifying some of *H. erectus*' early migrations out of Africa are as follows:

- GongWangLing (公王岭) in LanTian (蓝天) County, China: 2.12 millennia ago;
- Riwat in Pakistan: ~2 millennia ago;
- Dmanisi in the Caucasus: 1.81 millennia ago;
- YuanMou (元谋), China, two *H. erectus* incisors: 1.7 millennia ago;
- MaJuanGou III (马圈沟) and ShangShaZui (上沙嘴) in the NiHeWan (你喝完) basin, North China: 1.6–1.7 millennia ago;
- LanTian County (蓝天), China, a cranium: 1.63 millennia ago;
- Java in South East Asia: 1.7 millennia ago;

(a) (b)

Figure 2.9 Stone implements from Olduvai Gorge used during the lower Palaeolithic period, 2.9–1.7 million years ago by ancient Hominins. (a) Olduvai Gorge stone. (b) Oldowan-tradition stone chopper.

- Ubeidiya in the Levant: 1.5 millennia ago;
- XiaoChangLiang (小长梁), China: 1.3 millennia ago.
- XiHouDu (西侯渡) archaeological site in ShanXi (山西) Province, earliest recorded use of fire by *H. erectus*: 1.27 millennia ago.
- Western Europe: 1.2 millennia ago.

From the earlier information, *H. erectus* had dispersed from Africa throughout most of the Eurasian continent, reaching in the East as far as China and Southeast Asia and over a land bridge between Asia and the Americas, into North America and then down into South America.

These distributions have been traced through the identification of Oldowan lithic industry artefacts, a term taken from evidence at the site of Olduvai Gorge in Tanzania, Africa, where the archaeologist Louis Leakey first discovered in the 1930s, Oldowan stone tools, Figure 2.9(a) and (b).

This brief description has not included the more recent migrations, 15,000 years ago, from East Asia into North and South America via the land bridge and the human colonisation, 4,500 years ago, into present-day Northeast Canada.

As these journeys were made on different continents by a gradually evolving mankind, knowledge was being accumulated from these environments, with their differing temperatures, humidities, plant and animal species. That knowledge has been accumulated and much has been passed onto their modern human descendants, although no doubt much knowledge has also been lost.

2.4.2.2 Natural environment

The natural environment that humans were exploiting has been that shown in Figure 2.4, influenced by earlier continental connections, Figure 2.5, and by Earth's temperature variations, Figure 2.6.

An important human impact has been the steady exploration of the natural environment from Europe to Asia to the Far East and eventually to the Americas.

From the fourteenth century to nineteenth century, there was a period of exploration from the Old World, Europe, to the New World in the Americas. This also led to the discovery, spread and development of important and valuable new foodstuffs.

2.4.2.3 Use of fire and tools

The gradual development of hand tools, first from stone and then from a widening range of metals, led to the ongoing human exploitation of their environments. At first this was through hunting–gathering but then farming and has been applied to an increasing technological exploitation of water, minerals, oils, gases and other materials, some of which will be described later.

2.4.2.4 Geography

Scientists have suggested that *H. sapiens* in Africa developed through a merging of populations from North, East and Southern Africa.

- The fossil records confirm that *H. sapiens* 'anatomically modern humans', were living in Africa c 350,000–260,000 years ago;
- The earliest remains, including *H. sapiens* fossils, are from the Jebel Irhoud, Morocco, c 315,000 years ago;
- The Florisbad Skull is from South Africa, c 259,000 years ago;
- The Omo remains are from Ethiopia, c 233,000 years ago.

Around 65–50,000 years ago humans expanded out of Africa, launching the colonisation of the rest of the Earth by modern human beings.

Initially small numbers migrated out of Africa, maybe a few hundred or a few thousand, bringing with them their skills, tools, eating habits and perhaps some of their animal companions.

We believe that the first migrations journeyed up the East coast of Africa along the Red Sea and round into Arabia.

However, some scholars believe that some migrations may have been on primitive rafts from Africa across the Red Sea directly into what is now Yemen.

The first human groups to leave Africa, gradually modified their eating and hunting practices, finding new food sources, and learnt new skills appropriate to their locations in Arabia and the Middle East that they migrated to.

We also know, from the later spread of religions, Judaism, Christianity and Islam, that there was subsequently significant human traffic back and forth between Arabia and Africa over the last 4,000–5,000 years, for some of which there have been records.

Refer to Figure 1.2 for the map of human colonisation of the world from out of Africa.

2.4.2.5 Changing human agriculture, vegetables

Founder crops or primary domesticates were a group of flowering plants domesticated by early human farming communities in Southwest Asia and went on to form the basis of agricultural economies across Eurasia.

They were originally defined by Daniel Zohary and Maria Hopf as consisting of eight plant species domesticated by early Neolithic farming communities in the Fertile Crescent of Southwest Asia and went on to form the basis of agricultural economies across much of Eurasia, Southwest Asia, South Asia, Europe and North Africa.

The founder crops consisted of the following:

- Three cereals, emmer wheat, einkorn wheat and barley;
- Four pulses lentil, pea, chickpea, bitter vetch;
- Flax cultivated as a food and fibre crop in regions of the world with temperate climates.

These were amongst the first domesticated plants in the world. These founder crops were domesticated in the Pre-Pottery Neolithic period between 10,500 and 7,500 years ago.

Different species formed the basis of early agricultural economies in other centres of domestication. For example, rice was first cultivated in the Yangtze River (長江) basin of East Asia in the early Neolithic period. Sorghum was widely cultivated in sub-Saharan Africa during the early Neolithic, while peanuts, squash and cassava were domesticated in the Americas.

Notable among the crops that became worldwide staples to humans are the ubiquitous New World crops known as the Three Sisters; see Figure 2.10:

- Maize or corn;
- Climbing beans;
- Winter squash or pumpkins.

These New World crops were those that were native to the Americas, Figure 2.11, and unknown to the Old World before Columbus' voyage from Europe in 1492 AD.

Many of these New World crops are now grown around the world and have gradually become an integral part of the cuisines of previous Old World cultures.

2.4.3 Energy usage

2.4.3.1 Plants

These human migrations led to humans transporting some staple crops as they moved.

But the migrating groups will also have learnt how to exploit new plants in their new locations, altering their methods of hunting, feeding and structure building.

Figure 2.10 shows some of the key crops that developed to become staples of human nutrition.

Modern international food experience has shown how rapidly food knowledge, information and availability can spread from country to country.

The human food industry has become a complex, global network of diverse businesses supplying most of the food consumed by the world's population.

2.4.3.2 Animals

Animals became tamed as human communities developed and expanded. Probably the first domestic animal being the dog 'man's best friend'.

Figure 2.10 Three vegetables exploited for food by humans, the Three Sisters. (a) Maize or corn; (b) climbing beans; and (c) winter squash or pumpkin.

It seems likely that domestic dogs accompanied humans as they migrated to new locations out of Africa and perhaps some other domestic animals were acquired, including pigs, chickens and cattle for food, then donkeys and horses as stronger work, for hauling ploughs and carts and providing early forms of transportation.

2.4.3.3 Human understanding

Humans developed the ability to exploit fire, probably learnt from observing bush fires as a result of lightning strikes to trees or brush.

They then used fire to clear lands for planting, then for heating and cooking the plants that they had cultivated or gathered.

Humans used their own ability to haul produce and developed the wheel and the cart to transport foodstuffs and materials. Once the cart was developed, humans

Figure 2.11 America's Discovery, viewed by Native Americans, painting Thomas Hart Benton, 1922

could harness an animal to it, for example a dog, donkey or horse to haul progressively larger loads.

Then later larger wheeled carts appeared and ploughs were used to till the land. These carts then gave humans the ability to transport produce and people over significant distances.

Large rivers in Africa also facilitated riverine trade between communities, using at first small log rafts, skin boats, coracles or possibly bark boats; see Greenhill and Morrison [13].

The Nile in Africa became one of the first important highways for trade, a forerunner of all the Earth's great trading rivers; see Figure 5.1.

Between Africa and Arabia, the presence of relatively calm seas also facilitated travel by boat and increasing trade, stimulating the development of more substantial oared rowing boats and then sail-powered vessels.

There must have been a gradual human realisation of the social benefits from exploiting their own animals' energetic capabilities as well as the value of fire to heat water, burn scrub and manipulate stone, wood and other materials.

This is the very earliest human understanding of the role of energy and power, possibly 10,000 years ago.

2.5 Summary

This chapter has shown the changes that took place on Earth from 542 million years ago to the start of the Historic Past.

What is initially described took place before the Earth's human populations started.

The chapter demonstrates the huge, non-human effects of the interaction of the Earth's vulcanism and continental drift, which shaped our present continents, altering their nature and geography. This resulted in the consequent extinctions of many plants and animal species before humans emerged. It also implies the huge future changes that the Earth may undergo, with or without human influence.

The chapter explains some of the impacts of geology, plants and animals upon the Earth's environment, before humankind, changing dramatically the Earth's geography, animal and plant populations during this period.

The Earth has been in continuous geological and meteorological turmoil during the period, which has been part of the development of the Earth's subsequent life and climate.

Plant and animal species extinctions led to the deposits of future fuels, which early humans subsequently exploited. The coherent structure of the Earth and its development has now become much clearer to us.

Then came the development of today's recognisable Earth geography, its consequent plants, animals and the human race. However, the entrance of humans has altered this dynamic due to the energetic human exploitation described in this book, further influencing the causal effects of climate change. This is now leading to a greater human understanding and concern for our impact on our environment.

The challenge for future chapters is to unravel how humans can change their activities to protect the Earth and to be able to live in partnership with our environment.

Chapter 3
Historic past, last 5,000 years

3.1 Introduction

The evolution of the ability of *Homo sapiens* to extract energy from their environment over the last 5,000 years is a complex, intricate and specialist subject, but it can be summarised briefly here, before considering the most recent 2,000 years.

Initial efforts for humans, over many years, were the hunting of prey for defence and food, followed by the growing of crops, the use of fire and more recently the exploitation of watercourses and the sea. These efforts were initially sparse and in open locations in the eastern and central parts of the African continent, gradually moving into the Eurasian continent, down to Southeast Asia and into present-day Philippians, Indonesia, latterly Australasia and finally over the Bering Sea into North and then South America.

With the development of tools, humans mastered explorations of the land and the culling and subsequent cultivation of the following:

- Plants and fungi for food and medicines;
- Trees for firewood and cooking, then the use of boughs and trunks for specific home, exploratory navigation and then larger structural needs;
- The exploitation of rivers and seas for fish for food;
- From mastering these skills and technologies, many humans formed isolated, cultivated areas, development of simple small farms, followed by the creation of more sophisticated homesteads, surrounded by cultivated land and villages and ultimately the development of small townships;
- Some human tribes, however, for example in Africa, New Guinea and Australasia, in wide sparse plains, continued to maintain a peripatetic lifestyle, travelling and searching for food but not forming fixed settled communities, subsequently also repeated in the Americas;
- During these processes some collaborations developed between nearby groups or tribes but also some conflicts between them.

Finally, humans used their ability to travel by land, river and sea to begin to explore and exploit the known world, to simply to avoid their enemies, animal or human expanding their reach and access to resources.

This chapter is largely about methods of extracting energy for human use and suggests how the earlier-mentioned processes lead humans to extract energy from the Earth's environment and started to learn the consequences to their environment of these activities, before the surge in use of hydrocarbons.

3.2 Sun and Earth

3.2.1 Sun

The Sun, a nearly perfect ball of hot plasma, heated to incandescence by nuclear fusion reactions in its core, is the star at the centre of our solar system. It radiates energy as light, ultraviolet and infrared radiation and is the most important source of energy for life on Earth.

The Sun is the source of energy for our solar system, see Figure 3.1, which shows:

- The Sun's basic core structure, the core, radiative zone, convective zone, photosphere and chromosphere;
- Features of the Sun include the following:
  - Sunspots;
  - The corona;
  - Flares;
  - Prominences.
- And the presence of a solar wind;

The energy from the Sun has powered all life on Earth and sustains all our energy extraction systems.

Naturally the Sun cannot have a limitless life so there must be a limit to all life on Earth.

The Sun has a radius of about 695,000 km, or 109 times that of Earth. Its mass is about 330,000 times that of Earth, comprising about 99.86% of the total mass of

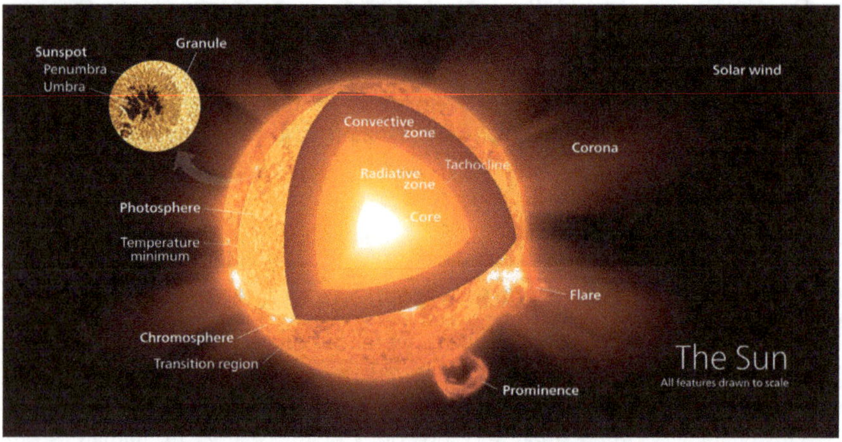

Figure 3.1 The Sun. Attribution: Kelvinsong

the solar system. Three-quarters of the Sun's mass consists of hydrogen (~73%). The rest is mostly helium (~25%), with much smaller quantities of heavier elements, including oxygen, carbon, neon and iron.

The Sun's role, in terms of this book, is as an Energy Vector and life-giver to Earth.

3.2.2 Earth's winds

Human developments and maritime exploitation have been stimulated, not only by human innovation, but also by the vagaries and predictabilities of the winds, as shown in Figure 3.2, circulating around the coasts and rivers of the populated parts of the African, Asian, European and American continents.

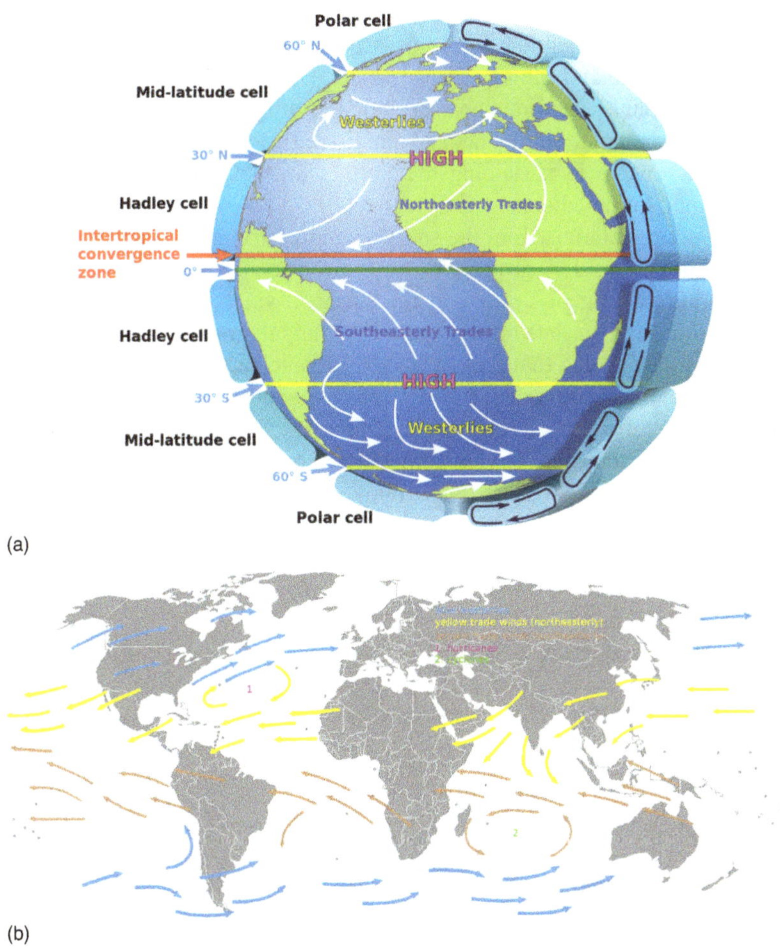

Figure 3.2 Earth's winds. (a) Earth's wind mechanisms and (b) Earth wind map.

These winds have increased the rate of exploitation, particularly by Europeans of travel to the Americas, Africa, India and the Far East, from the fourteenth century onwards.

However, modern scholarship has now brought to light the extensive human exploration from Southeast Asia into the Pacific and the colonisation of Pacific Islands.

This developed over 2,800 years from 2000 BCE to 800 CE.

This then included the far colonisation by 1500 CE of New Zealand, or Aotearoa in the Maori language.

3.3 Energy usage, BCE

3.3.1 Energy extraction mechanisms

3.3.1.1 Plants and fungi

Historically, the plant kingdom encompassed all living things that were not animals, including algae and fungi.

All current plant definitions now exclude the fungi and some of the algae.

We now understand that plants and fungi are different but have an important symbiotic relationship in sustaining the Earth's present life and are valuable to humans for food and medicines.

Fungi extract materials from the soil exchanging some of those materials with plant roots. Fungi spread by growing extensively, laterally through the soil. Fungal structures grow fruiting bodies above the surface, which burst and spread fungi spores over the soil's surface to propagate new fungal structures.

Early man discovered that fungi are useful medicines. Medicinal fungi contain metabolites, which can be induced to produce metabolites, through biotechnology, for the development of prescription drugs. Compounds of fungal origin successfully developed include some antibiotics, anti-cancer drugs, cholesterol and ergosterol synthesis inhibitors, psychotropic drugs, immunosuppressants and fungicides.

Plants on the other hand extract nutrients from the soil via their roots, exchanging materials through filamentary fungal structures, using energy extracted from the sunlight falling on their leaves. Each plant is a complex network connecting the roots to its leaves and flowers.

Many plants propagate sexually, using flowers to attract insects, by colour and scent, spreading male spores on the insect body, which are then carried by the insect to fertilise other female spores in the flowers of the same or nearby plants, of the same species.

3.3.1.2 Insects, animals and humans

In Figure 3.3(a) an insect carries off in its mouth a cut leaf from a plant. These leaf parts are returned to the ants' nests where the chewed leaf material becomes a bed for the growth of fungal food to sustain the ants' hives.

Insects also prey on other animals, for example in Figure 3.3(b), where a mosquito is feeding off a human, by inserting its proboscis into the skin and sucking blood as food.

Figure 3.3 Insects, animals and humans, food and energy. (a) Ant with a cut plant leaf returning to its nest for processing as fungal food; (b) mosquito extracts blood as food and energy from human skin; (c) plaster cast of a large ants' nest colony; (d) intensive modern battery chicken feeding and egg-laying; and (e) intensive modern pig feeding.

Rather similar intensive human farming of poultry and mammals for food can be seen in Figure 3.3(c) and (d).

The latter two pictures may give the reader a rather unsatisfactory view of human exploitation of the animal kingdom. However, Figure 3.3(e) demonstrates that large insect nest structures are no less complex nor less exploitative than human food structures.

3.4 Human energy usage, 5000 BCE to 2000 CE

3.4.1 Human engineers

This section is a description of our historic past, including human engineering influences on the Earth's environment, by geographical location, and the changes caused during that period with observations on how that has changed as human settlements expanded, during the recent past, from 1 to 1850 CE.

This sets the scene for later descriptions of the more up-to-date use of energy by humans, up to the beginning of the modern era and the recent huge modern expansion of human energy usage.

3.4.2 Earth's surface

3.4.2.1 Wood, grass, plants and settlement

Early man exploited the surrounding physical environment for wood to make fire and cook, for grass to feed livestock and plants for food.

This process of domestication and exploitation started in Africa but then expanded as people migrated out of Africa. The varieties of animals domesticated expanded, as new territories were occupied.

The sizes of land plots increased as human populations and domesticated animal herds grew. But land plots varied according to the customs and sizes of human family or tribal groupings changed.

There are still small rural human settlements, particularly in Central Asia and Africa, for example as shown in Figure 3.4(a). However, many humans now live in huge, industrialised cities.

The author has been a regular academic visitor to the main University in Chongqing (重庆) China, Figure 3.4(b).

This city is a huge, modern human settlement, with a current population of 22.5 million people.

The city played a major role in China's opposition to the Japanese following their invasion of China in 1931. The city became the headquarters of the resistance of Chiang Kai Shek's Guómíndǎng Government, during the later stages of Japan's occupation of China.

Following the establishment in China of the communist Gòngchǎndǎng Government, Chongqing became an important Third Front base for industrial production, including the development of defensive industries.

Chongqing's present prosperity is based upon those earlier developments; however, see Jackson [23], the city's original prosperity was based upon its site on the Yangtze River and an ancient Chinese dynastic settlement dating back 2,400 years.

Humans are attracted to cities where wages can be higher, but accommodation is costly, and it takes time in a city to build up the social relations that came so easily in smaller rural settlements.

3.4.2.2 Food

The modern food industry has become highly diversified, with manufacturing ranging from small, traditional, labour-intensive family-run activities, as in

Figure 3.4 Human settlement. (a) Small South African Ndebele settlement, 1999, and (b) huge Chinese settlement on Yangtze River, Chongqing (重庆), 2010.

Figure 3.4(a) to huge, capital-intensive, highly mechanised industrial processes, as in a big city like Figure 3.4(b).

On the other hand, many parts of the food industry around the world, even supplying big cities, still depend largely upon local agricultural produce, animal farms and fishing.

It is difficult to find an inclusive way to describe all aspects of an up-to-date food production and sales industries.

- The UK Food Standards Agency describes it as:

  A whole UK food industry, from farming and food production, packaging and distribution to retail and catering.

- The Economic Research Service of the USDA uses the term food system to describe it as:

  The US food system is a complex network of farmers and the industries that link to them. Those links include makers of farm equipment and chemicals as well as firms that provide services to agribusinesses, such as providers of transportation and financial services. The system also includes the food marketing industries that link farms to consumers, and which include food and fibre processors, wholesalers, retailers, and foodservice establishments.

Modern food industries include the following:

- Agriculture:
  Raising crops, livestock, and seafood and agricultural economics;
- Manufacturing:
  Agrichemicals, agricultural construction, farm machinery and supplies and seed;
- Food processing:
  Preparation of fresh products for market, and manufacture of prepared food products;
- Marketing:
  Promotion of generic products (e.g., milk board), new products, advertising, marketing campaigns, packaging, public relations;
- Wholesale food distribution:
  Logistics, transportation and warehousing;
- Food services, including catering;
- Grocery, farmers' markets, public markets and other retailing;
- Regulation:
  Local, regional, national, and international rules and regulations for food production and sale, including food quality, food security, food safety, marketing/advertising and industry lobbying activities;
- Research and development:
  Food science, food microbiology, food technology, food chemistry and food engineering;
- Education:
  Academic, consultancy, vocational;
- Financial services:
  Credit and insurance.

3.4.2.3 Early pottery, implements and tools

In China the development of colourful and sophisticated pottery bowls was occurring 6,000 years ago, around the central basin of the country.

An example is shown in Figure 3.5.

This is from the YangShao culture, referring to the Neolithic period along the Yellow River (黄河) in the provinces of Henan, Shaanxi, Shanxi and Gansu but also stretching further into neighbouring provinces, including Szechuan.

Figure 3.5 *Decorated YangShao culture clay-fired bowl of the Miaodigou period, 4000–3500 BCE, China. Courtesy: Dr GuoHua Yang.*

The earliest tools used by humans to develop the Earth's interior were sticks and probably the horns of hunted animals.

A historical vignette, close to the author's home, is one of the earliest Neolithic, 4300–2000 BCE, mining sites in the British Isles at Grime's Graves in Norfolk. This is an area of chalky ground with the only Neolithic flint mine in Britain currently open to visitors.

The area, Figure 3.6(a), has a grassy, almost lunar landscape of 400 pits, being named in Anglo-Saxon times as Grim's Graves, 1,400 years ago.

It was not until the 1870s, when one of these, so-called graves, was excavated that they were identified as flint mines, originally dug into a chalk subsoil over 5,000 years ago, initially to extract flints for toolmaking and construction.

The tools used by early East Anglian Neolithic inhabitants, to excavate these shallow mines, were originally manufactured from deer's antlers, and were surprisingly intricate, Figure 3.6(b). They were used to extract flint nodules from the chalk and that flint was then split and cut to be used to produce projectile points for spears, scrapers, knives and other stone tools.

Activity continued at this site up until the eighteenth century, by which time the flint was also being mined primarily for use as building material and used in nearby Norwich's fifteenth century Guildhall, Figure 3.6(c).

A more recent subsequent application was for the manufacture of flints to provide the spark needed to ignite gunpowder and fire a gun.

3.4.2.4 Water in the land

China is a country where its large rivers played a huge part in human development.

Figure 3.7 shows China's current extent, locating its two main rivers, the Yellow River, Huang He, (黄河) to the North, and Yangtze River, Cháng Jiāng (長江) to the South.

The map at Figure 3.7 also shows the location of 15 selected sites where YangShaoWenhu (仰韶文化) culture pots, have been found, as depicted in Figure 3.7, between the great Yellow (黄河) and Yangtze (扬子江) rivers.

3.4.2.5 An early water irrigation scheme

Dujiangyan (都江堰) was an early irrigation system on the River Min, Mínjiāng, a lengthy tributary to the larger Yangtze River (長江) in Sichuan (四川), China, as shown in Figure 3.8. It was originally constructed in 256 BC by the State of Qin, an area in the western part of the Chengdu Plain, between the Sichuan Basin and Tibetan Plateau.

Originally, the Min descended rapidly from the Min Mountains, slowing down abruptly when reaching the Chengdu Plain, a process which filled watercourses with silt, making the surrounding fields prone to flooding.

In 260 BC, King Zhao of the Qin commissioned a project to construct the Dujiangyan flood-control system, harnessing the river by using a novel method of division and channelling, to slow down flowrates, rather than by simply damming the river.

Figure 3.6 *Grime's Graves, Norfolk, United Kingdom, site of extensive flint mines from Neolithic times 3000 BCE to the eighteenth century. (a) Dating from 3000 BCE, extensive mining for flints initially for tools. (b) Antler tools for extracting flints a typical early flint blade (c) Norwich fifteenth-century Guildhall built and decorated with knapped flints. Courtesy: English Heritage.*

That water management scheme is still in use today, irrigating over 5,300 square km of land in Sichuan Province and providing comprehensive agricultural benefits in terms of flood control, irrigation, water transport and better water management, and Dujiangyan has become a scenic tourist area.

Figure 3.7 China, its two main rivers, Yellow to the North, Yangtze to the South, early pottery locations in the centre. Courtesy: Dr GuoHua Yang.

3.4.2.6 Seas and human explorers

Exploration of the seas has been a very important aspect of the spread of humans worldwide and this exploration has become established in a number of specific locations where the seas' violence is moderated by the coasts' geographical arrangement.

Particular locations where early human sea-traffic flourished were as follows:

- The Red Sea, the coast of Arabia and the Persian Gulf;
- In Europe the Mediterranean Sea and in near Asia the Black and Caspian Seas;
- The South China Sea and the coasts of South China, Vietnam, Malaysia and Indonesia;
- The North, Baltic and Norwegian Seas and the North Atlantic Ocean between Scandinavia, Iceland, Greenland and North America.

These areas formed the cradle of human sea navigation but those developments were first stimulated by human riverine trading which debouched into these from the world's great rivers in China, SE Asia, Europe, Arabia and the Indian subcontinent.

The exploration and mapping of the seas and coasts provided a huge boost to these trades, pioneered by the following but non-exhaustive list of explorers and naval officers from many countries on Earth:

- From China:
  Xu Fu, Zheng He, Chang Chun and Du Huan;
- From Portugal:
  Afonso de Albuquerque, Ferdinand Magellan, Vasco da Gama and Francisco de Almeida;

(a)

(b)

Figure 3.8 *Early river-channelling for advanced flood control and land irrigation, Dujiangyan City, Sichuan, China. (a) Present-day view of the Min River bend at Dujiangyan and (b) map of the river control system.*

- From Spain:
  Pedro de Alvarado, Christopher Columbus, Juan Bautista de Anza, Lucas Vázquez de Ayllón and Vasco Núñez de Balboa;
- From Denmark:
  Vitus Bering and Jørgen Jørgensen;
- From the United Kingdom:

Henry Hudson, Richard Hawkins, Thomas Cavendish, Francis Drake, William Baffin, George Anson, James Cook, Isabella Bird, Mary Henrietta Kingsley and James Clark Ross;

- From the Netherlands:
  Abel Tasman, Willem Schouten, Cornelis de Houtman and Maarten Gerritsz Vries;
- From Russia:
  Alexander Kolchak, Maria Pronchishcheva, Eugene Smurgis and Grigori Ivanovitch Langsdorff;
- From the USA:
  Frederick Russell Burnham, Richard E. Byrd, Harriet Chalmers Adams and Edmund Fanning.

3.4.3 Modern water schemes

3.4.3.1 Plains and lakes

An ambitious river and lake scheme was developed by Egypt on the River Nile and includes the Aswan Low Dam and its later much larger scheme the Aswan High Dam, Figure 3.9(a).

The earliest recorded attempt to build a dam on the Nile near Aswan was in the eleventh century when the Arab engineer and polymath Ibn Il-Haytham was ordered by the Caliph to regulate the flooding of the Nile.

The modern High Dam, built between 1960 and 1970 across the River Nile in southern Egypt near the city of Aswan, is 3,830 m long and 111 m high, 980 m wide at the base, tapering to 40 m wide at its crest; the structure contains 43,000,000 m^3 of rock and clay materials.

It was designed, like its predecessor, the Low Dam, not only to control the River Nile flooding but also to supply electricity to a growing Egyptian economy.

Egypt has relied for thousands of years on the river's annual seasonal floods for both drinking water and farmland irrigation.

The High Dam replaced the Old Aswan Low Dam, further downstream, 1,950 m long and 22 m high, built in 1902 by British engineers and Egyptian workers.

The Low Dam had a navigation lock on its western bank to allow ships to pass through and when it was built the Low Dam was the largest masonry dam in the world, constructed from locally sourced Aswan granite.

Engineers built the Low Dam in 1902 with 1,802 m width of under-sluices at its base. These channels allowed water and silt to pass through the dam when the Nile flooded, preventing the upstream reservoir from silting up. The expanding Egyptian cotton industry and its need, between 1910 and 1933, for water required the Low Dam to be raised twice, additions which brought its height up to 36 m, boosting to 58 m^3 the amount of water it could hold.

Egypt's growing population and the Egyptian government's modernisation plans had, by the 1950s, rendered the Low Dam less effective. There were also anxieties that the Low Dam's higher wall could result in upstream flooding of the important ancient Philae temple. Plans were then drawn up for a second High Dam, further upstream.

(a)

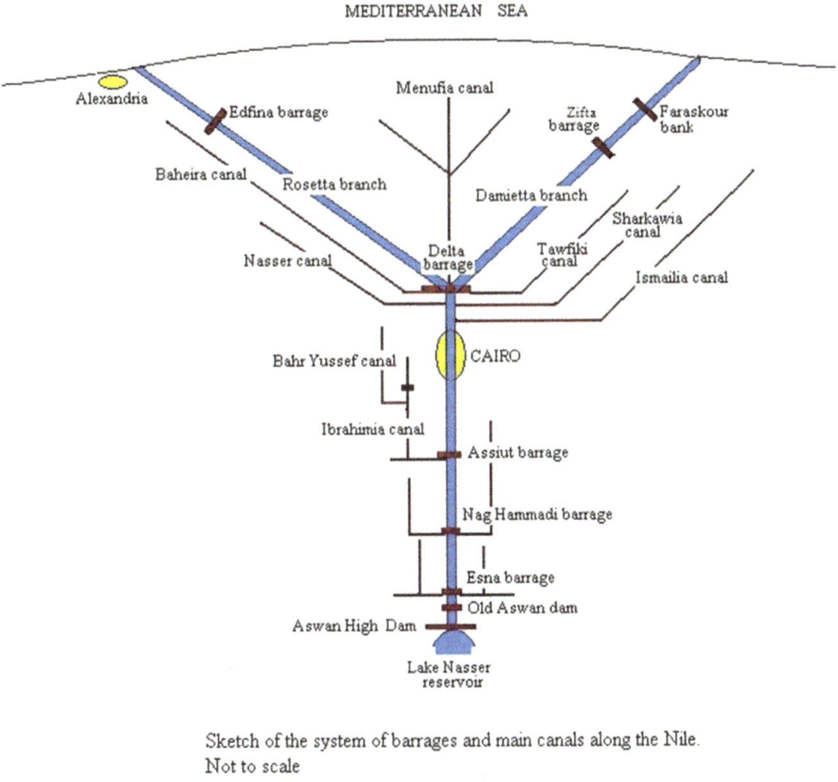

(b)

Figure 3.9 Flood control schemes present status, showing Aswan High and Low (Old) Dams and their downstream influence. (a) Aswan High Dam viewed from space. (b) Scheme of River Nile from the High Dam to Mediterranean Sea.

The High Dam was designed to impound Lake Nasser, named after the Egyptian President being built between 1960 and 1970 across the River Nile in southern Egypt with key details as follows:

- A structure of 3,830 m long, 111 m high above its base, 980 m wide at its base, tapering to 40 m wide at its crest;
- The structure contains 43 million m^3 of rock and clay materials;
- The dam reservoir is on average 550 km in length and up to 35 km containing 132 km^3 of water, compared to the 4.9 km^3 holding of the Low Dam, with a total surface area of 5,250 km^2;
- The water level can be at a maximum of 185 m above sea level, 130 m higher than the river level downstream;
- Peak power output 2,100 MW.

The High Dam's purpose was for more effective control of Nile floods, more water for hydroelectric generation and for irrigation, factors crucial to Egypt's industrialisation plans.

The High Dam was funded by a loan of $1.1bn from the Soviet Union to Egypt, equivalent today to £11.5bn, granted at that time as the Soviet Union was increasing its influence in the Middle East.

Soviet engineers in the Moscow-based Hydro Project Institute designed the scheme in Russia. The project team in Egypt built a series of emergency spillways or overflow channels into the High Dam to deal with increased volumes of water that future heavy rainfall, further upstream, might bring.

Soviet and Egyptian engineers and Egyptian workers undertook this huge Project, relocating more than 90,000 people from areas the High Dam was set to flood.

Some were removed from their original homes as far as 45 km; similar removals were also necessary in China during the development of the Three Gorges Dam project, SanJiaDaBa (三峡大坝), in China, which will be described in a later chapter.

Around 25,000 Egyptian engineers and other workers played a part in the scheme. The full downstream impact of Egypt's Nile River management can be seen in Figure 3.9(b), showing evidence for the huge effect the project has had on Egypt from the High Dam.

3.4.3.2 Rivers

A significant source of energy that can be abstracted from the environment can be found in the water flowing in rivers or drawn from the damming of rivers.

An early example of such a water management scheme was that installed nearly 2,300 years ago, at Dujiangyan in Sichuan Province, China, described in Section 3.4.2.6, still in use today.

The origin of these schemes is initially the avoidance of flooding, but such human interventions can also be developed to extract energy.

A more modest example of flood control on the River Thames in United Kingdom is shown in Figure 3.10.

Problems arise because the civil engineering costs necessary for energy extraction mitigate against flooding limitations.

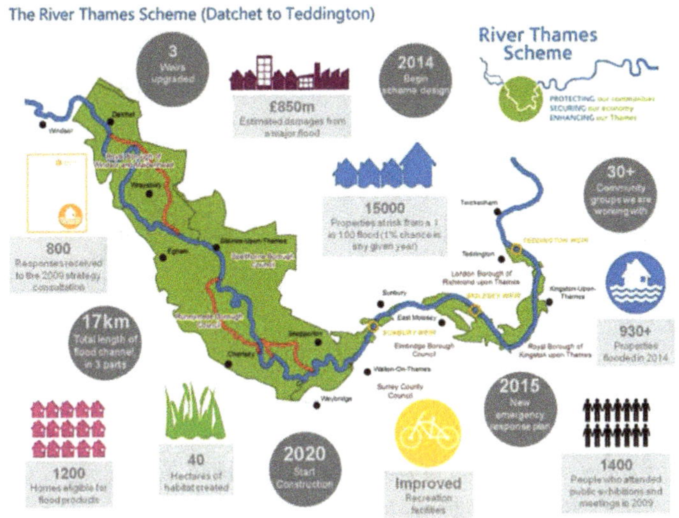

Figure 3.10 Current River Thames flood control scheme, United Kingdom

However, the present accumulated world experience of flood control, combined with energy extraction, means that there is now an opportunity to develop more ambitious schemes.

That completed in 2012 by Chinese engineers and workers on the Yangtze River in China at the Three Gorges Dam, SanJiaDaBa, is impressive and rather a dauntingly large example, to be described in a later chapter. However, an earlier river and lake scheme was developed in Africa, in Egypt and its story will be told in the next section.

3.4.4 Earth's interior

3.4.4.1 Coal, oil and gas

Coal, oil, tars and gas have coexisted in Earth's geological measures for millennia, having been laid down by millions of years of plant or animal growth and death, being embedded in the Earth's crust and distributed geographically by the effects of plate tectonics, continental drift and volcanic activity at the edges of the Earth's tectonic plates, Figure 3.11.

Coal has been extracted for centuries by humans from the Earth's surface for use as energy, initially for fires, for cooking and heating, but progressively in the last millennium for increasing industrial activities: the forging of metals, including copper, iron; the manufacture of steel and for similar purposes with more complex metals and alloys.

Coal is a combustible black or brownish-black sedimentary rock, formed as coal seam strata; it occurs in a variety of qualities ranging from the hard, energetic, shiny black anthracite found in West Wales to the low-grade light dusty lignite found in Northern Germany, Poland and Northern China.

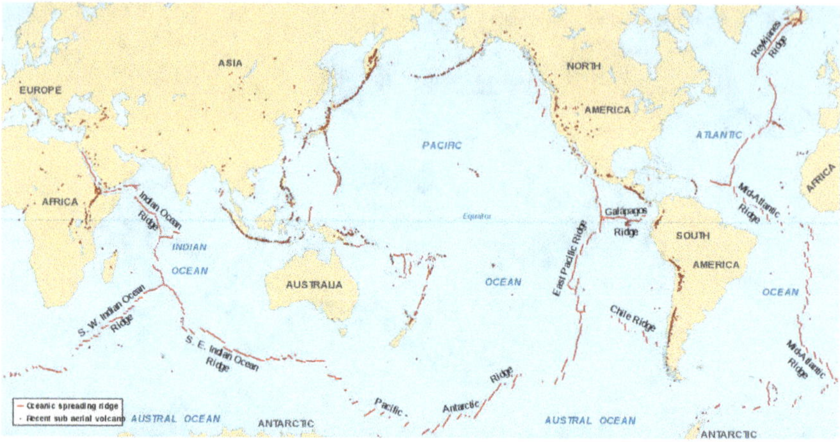

Figure 3.11 Current locations in red of tectonic plate separations, consequences of continental drift on the Earth's surface

Coal is mostly carbon with variable amounts of other elements, chiefly hydrogen, sulphur, oxygen and nitrogen.

By the fifteenth century there were attempts in many countries to extract tar from coal measures, for use in lighting and or waterproofing.

Fisher [35] gives a very comprehensive history of human discoveries of gas, naphtha, oils and tars, showing the wide range of regions and societies who made these discoveries, in all parts of the habitable Earth and the uses they put them to.

Oil and gas effluents from rocks, sands and muds had been reported between humans in diverse societies over more than 3,000 years, in many parts of the world.

There had been a realisation that these oil and gas resources were the result of effluents from compressed prior vegetative and animal materials, or hydrocarbons, seeping from concentrations in specific rock, sand or mud types. They became apparent to humans in the water of streams and rivers passing through these geological formations.

Many societies reported that, as these oils and tars were released from the surface, they proved useful for water-proofing boats and habitations or as medicines for various ailments. The gases from some such locations will have spontaneously ignited, due to lightning, volcanic activity or the intervention of humans or animals, so it was clear that these resources could be flammable and useful for lighting.

Visible surface features, such as oil and natural gas seeps, shallow or deep pockmarks or underwater craters, caused by escaping gases, provided basic evidence around the Earth of these hydrocarbon resources close to the surface.

The geographical spread of these locations was very wide, including all of Earth's main continents, with sites in the Americas, Ireland, Great Britain, Europe, Central Asia, Southeast Asia and China.

Modern maps with the distribution of these coal, oil and gas resources are shown in Figures 3.12 and 3.13.

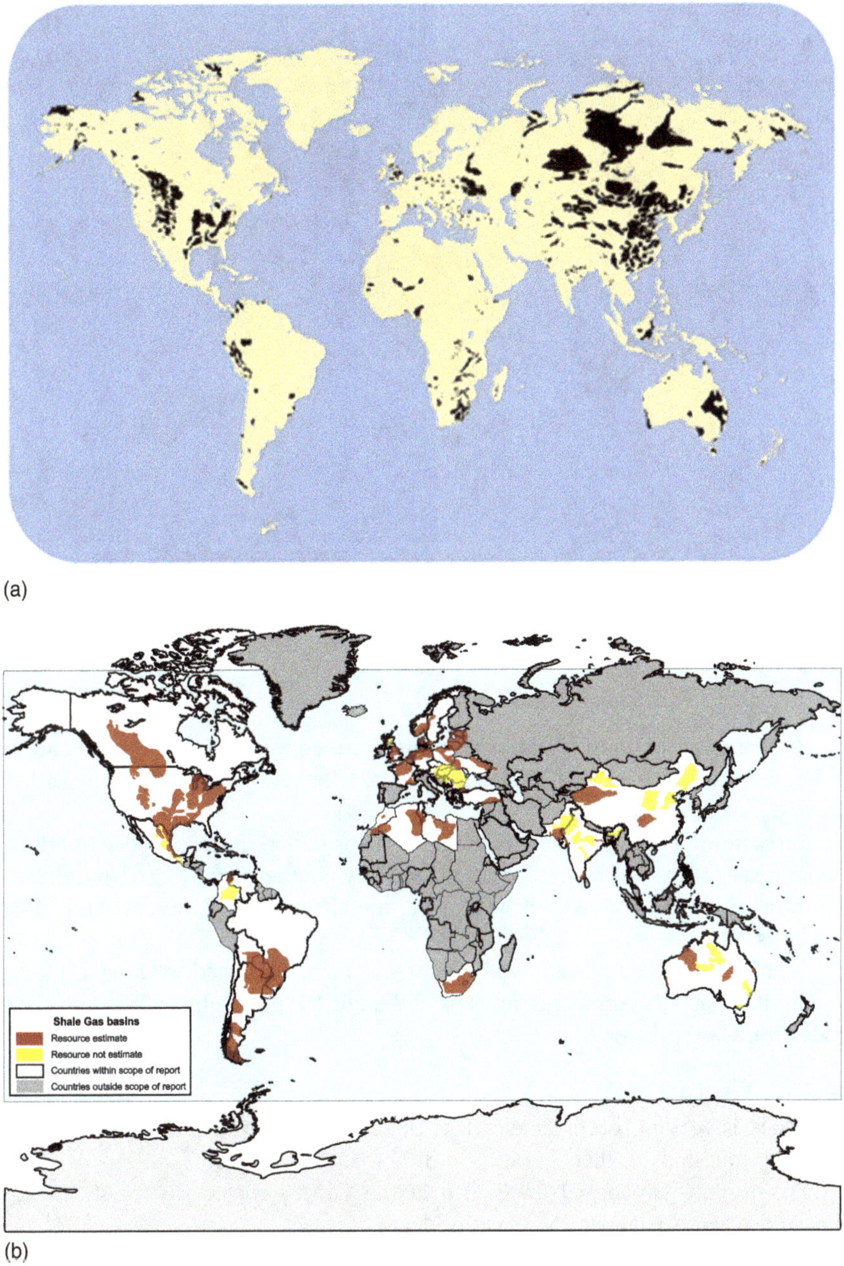

Figure 3.12 *Present assessed world coal, shale oil and gas fuel resources. (a) Coal resources and (b) shale oil and gas resources. Courtesy: World coal deposits and US Energy Information Administration.*

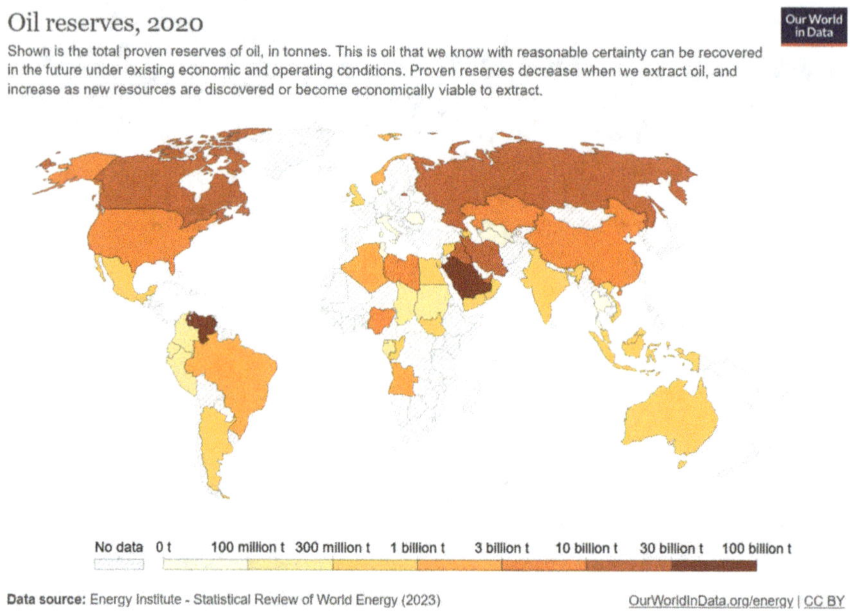

Figure 3.13 Assessed world oil resources. Courtesy: Our World in Data.

But some of the most significant deposits are concentrated in Europe and the Middle East, centred on what is now Romania, Ukraine, Southern Russia, Turkey, Iran and Saudi Arabia.

Early mining methods were crude, and the principal initial purpose of mining was to extract coal to burn, first domestically, but gradually coal became an important industrial power source for metallurgical, chemical and other industries.

Sophisticated methods allowed lighter oils to be extracted from coal, mainly for lighting but also some for heating, although by the eighteenth century that production was still small.

3.4.4.2 Geothermal

Volcanoes have long been imagined as potential energy sources, either through their heat emissions or their production of hot water and steam.

The normal geothermal gradient near the Earth's surface, or crust, in most parts of the world is about 25 °C/km of depth.

However, when volcanic magma, molten rock from the Earth's interior, enters that crust as a shallow intrusion beneath a volcano, Figure 3.14, the normal geothermal gradient is locally perturbed, and temperature rises around the intrusion.

Hot springs have been used for bathing since at least since Palaeolithic times.

The extent and duration of such thermal anomalies depends mostly on the temperature and volume of the intruded magma.

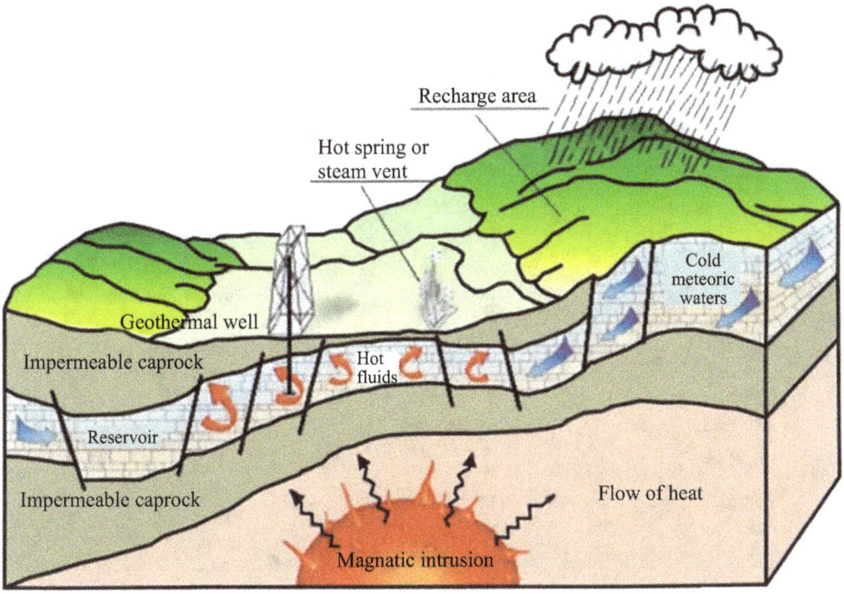

Figure 3.14 Exploitation of geothermal energy

The presence of hot magmas below the surface of active volcanic regions offers the prospect of harnessing significant amounts of geothermal energy. That energy is a renewable resource, as it exploits the abundant sources of Earth's interior heat and water, which, once used and cooled, can then be discharged back into its original reservoir.

The use of this natural energy source has important implications for preserving the environment and was recognised by earlier societies as a potential energy source. However, the presence of long-term stable energy source sites as shown in Figure 3.14 are relatively rare.

When the Romans conquered Britain in the first century CE they established the town of Aquae Sulis, now Bath in Somerset, United Kingdom. A temple was constructed on the site between 60 and 70 AD and exploited those hot springs there to feed public baths and the underfloor heating of nearby Roman residences. The admission fees to these baths probably represented the first commercial use of geothermal power.

The oldest recorded spa is at a site in China where the Huaqing Chi palace was later built, in 723 AD. HuaQing is a complex of hot springs located in an area characterised by mild weather and scenic views at the northern foot of Mount Li, one of the three major peaks of the Qinling Mountains 25 km east of Xi'an, formerly Chang'an, the western capital of the Tang dynasty, in the province of Shaanxi.

The world's oldest geothermal district heating system is in Chaudes-Aigues, France, and has been operating since the fifteenth century.

Figure 3.15 Current exploitation of geothermal hot springs, Larderello, Italy

The earliest industrial exploitation of geothermal energy began was at Larderello, Italy, in 1827 with the use of geyser steam to extract boric acid from volcanic mud.

This geothermal exploitation at Larderello continues today by ENEL, the Italian National Electricity Authority, generating electricity using this volcanic source, Figure 3.15.

3.5 Summary

This chapter has demonstrated early energy developments from 5000 BCE to 2000 CE in Europe, China and USA and elsewhere with some of their consequences, laying the foundations for energy exploitation elsewhere, described later in the chapter.

These developments demonstrate the universality of human energy demands in all the Earth's continents, initially for lighting but then for larger industrial use by means of coal, iron, oil, gas but also from waterpower.

The last sections of the chapter describe the appearance of the Coal Age and subsequent start of the Oil Age, their gradual and sometimes chaotic initiation in the eighteenth century and particularly the oil industry's catastrophic initial management during the nineteenth century.

Some of these issues were highlighted in Princess Victoria's eloquent description of their deleterious effects on the environment and people in the West Midlands of England. But this also emphasises their effect on advancing of industry and their subsequent expansion, which were economically beneficial, but not necessarily ecologically so.

Later chapters will now describe how energy exploitation has driven an improvement in scientific knowledge in their extraction methods and exploitation. This has led to a human appreciation not only of the benefits of energy exploitation but also of the risks inherent in them and the important issue of balancing these two factors.

Many of these developments, the so-called Industrial Revolution, were causing severe human and environmental damage, the initial effects of which, although isolated in various countries, continents and oceans, were initiating a public debate on the consequences of worldwide industrial expansion.

The author recalls a vignette, in 1965, returning to United Kingdom from the Persian Gulf while transiting the Red Sea serving in an RN frigate and reading '*The Silent Spring*' by Rachel Carson [10], sequel to '*The Sea Around Us*' [7], one of the first books to describe in detail the deleterious effects of humans on our environment.

This was clearly visible to me in that beautiful but overcrowded sea-lane in the Red Sea, with stunning views of the mountains of Arabia on the starboard hand and the rugged coast of Egypt on port hand.

The extraordinary twentieth-century clean energy development in Egypt of the Nile River control schemes using the Low and High Dams, delivered substantial benefits giving a modern example of what clean energy could be capable of, which will lead in a later chapter to China's huge Three Gorges Dam, SanJiaDaBa.

The Industrial Revolutions, concentrated in Great Britain, Ireland, Europe, North America and China, delivered enormous economic benefits due to the availability of powerful energy sources and growing trade, with a consequent expansion of industry and population. But these developments contained a warning of potential environmental damage that could affect animals, humans and the environment ensuing from the large-scale, worldwide exploitations of fossil fuel energy sources.

Chapter 4

Fuels, their science and commercial influence, 1800–1900 CE

4.1 Introduction

The purpose of this chapter is to demonstrate the significance of both scientific and commercial ideas to the changes in our fuel and energy usage in the nineteenth century, which marked the acceleration of fossil fuel extraction and the innovation around their varying processes, including an increasing understanding of their effects on the environment and the need for safety methods to prevent fatalities.

4.2 Coal, oil and gas

4.2.1 Facts and figures of light, heat and industrialisation

4.2.1.1 Coal

Despite the worldwide location of coal, oil and gas resources, shown in Figures 3.6 and 3.8, some of the most energetic initial exploitation of these resources occurred in Western Europe and North America in the mid-1800s, based upon the increased usage of coal, oil and gas in those locations in the nineteenth century, primarily for

- lighting,
- heating,
- early mass industrialisation.

A significant coal industry had been present in many parts of the world from Asia to Europe since the fourteenth and fifteenth centuries.

The following gives an insight into coal industry industrialisation, initiated first in the fifteenth century in China, then in the nineteenth century in Europe and the USA with sections summarising their effects on the environment.

Starting with some detailed facts and figures about coal, oil and gas, their compositions by percentage as received by weight, their hydrocarbon content and, in the case of coal, a diagram has been used to show the gases, chemical products and fuels, extractable from coal.

Table 4.1 shows the composition of a typical bituminous coal.

While Figure 4.1 shows the range of types of coal types available in different regions of the world, refer also to Figure 3.12(a).

Table 4.1 Composition of a bituminous coal, % as received by weight

Elements by weight	
Element	**Range (%)**
Carbon	63.8
Hydrogen	4.5
Nitrogen	1.3
Sulphur	2.5
Chlorine	0.3
Oxygen	6.9
Ash	9.7
Moisture	11.1

Those at the top of the figure are the hardest most energetic coals with the least ash, available for example in Great Britain, Ireland and the Donbas in Ukraine. These harder coals appear in deep underground seams and maybe the most difficult to locate and work.

Whereas those at the bottom of the figure with less carbon are less energetic with considerable consequential ash, available for example in China, Northern Germany and Poland but appear in larger sometimes exposed seams and prove easier to work and industrialise than deep-mined coal.

Figure 4.2 gives an indication of the very wide range of gases, chemical products and fuels extractable from coal, depending upon the coal type in Figure 4.1. Coal-gas was developed by William Murdoch at Ayr in Scotland.

An important aspect of the use of coal was its need for a fluid which could transport energy from the fossil fuel furnace to the engine providing the output power.

The most readily available resource for that task was water, converted to steam from burning the fuel in the boiler, then used to drive the engine that delivered the power and finally condensed back to water at the end of the process.

Experience showed that this water and steam were exceptionally important to the new types of engines being developed and that steam had to be managed to deliver reliable heat transfer.

On completion of the heat transfer, this steam was returned to water in large, costly condensers and then reused in the boiler. See Figure 5.12 for an example of the size of water-recovery condensers in a large steam ship. The water had to be pure, free of salts to avoid scale, corrosion or foaming in the boiler and, because of the cost these requirements, the water had to be retained and reused, again and again.

An important phase in the use of other fossil fuels, oil and gas was the development of engines that did not require complex water management systems, like the condenser. In the future, this led to the increasing importance of internal combustion engine technologies for motor cars, lorries and aircraft, using oil and gas fuels directly without cumbersome condensers.

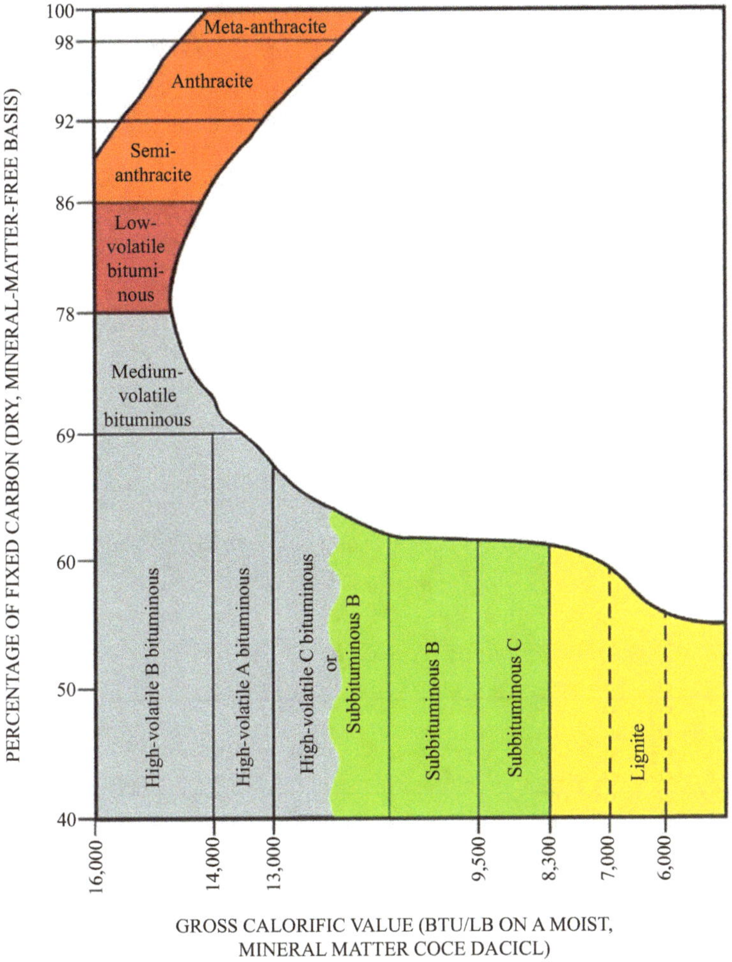

Figure 4.1 The quality of different grades of coal mined in different regions of the world

4.2.1.2 Oil and gas

Table 4.2 shows the composition of a typical oil and the light and heavy proportions by weight. The development next shows that early development of the use of oil as a fuel was initially of the lighter components for lighting.

But development of oil usage has shown that the heavier bitumen fractions of oil were widely used in historic for building in Iraq and Persi and for sealing the bodies of boats in many diverse countries and cultures.

The complex contents of these oils and gases also hint at their future potential for use as chemical sources for many materials, such as plastics, oils and fats, rather than as fuels; see Allwood and Cullen [25].

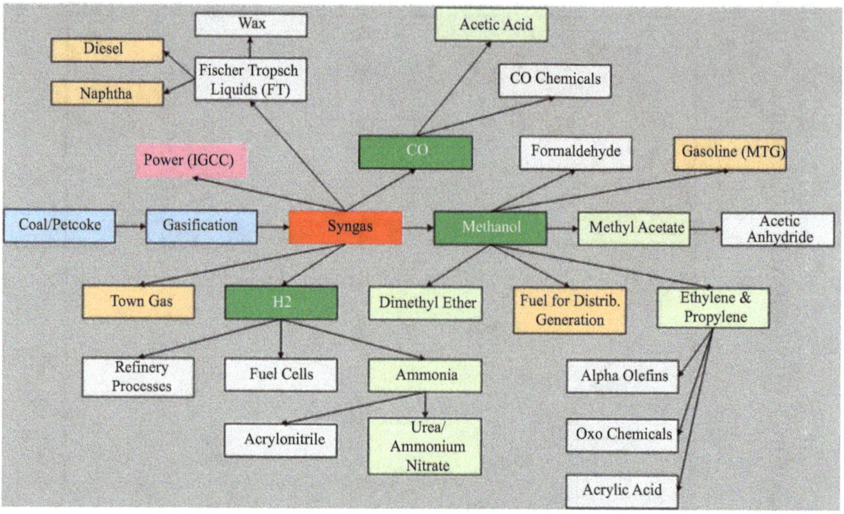

Figure 4.2 The gases, chemical products and fuels extractable from coal and coke

Table 4.2 Composition of oil, as received by weight, %

Elements by weight			
Element			**Range (%)**
Carbon			83–85
Hydrogen			10–14
Nitrogen			0.1–2
Oxygen			0.05–1.5
Sulphur			0.05–6.0
Metals			<0.1
Hydrocarbons by weight			
Hydrocarbon		**Average (%)**	**Range (%)**
Light	Alkanes, paraffins, including kerosene	30	15–60
Light medium	Naphthenes	49	30–60
Heavy medium	Aromatics	15	3–30
Heavy	Asphaltics, tars	6	Remainder

4.2.1.3 China coal developments

Figure 4.3(a) shows an early shallow coal mine in China during the Ming Dynasty 500 years ago, whilst Figure 4.3(b) shows an active coal mine in Szechuan Province in 1944, as the then Guómíndǎng government fought back against the Japanese invasion of China.

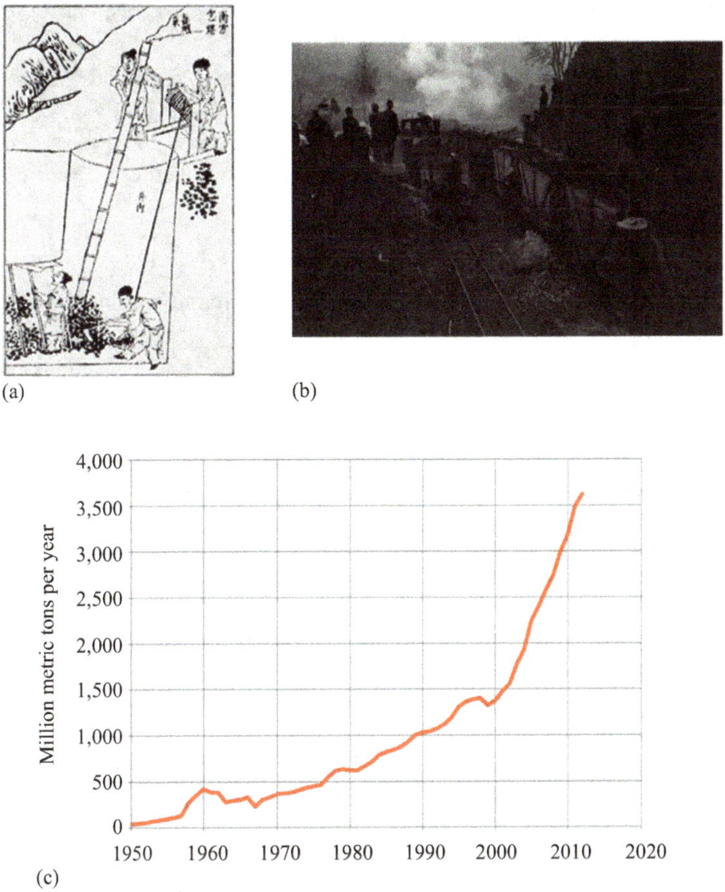

Figure 4.3 *Coal developments, China. (a) Ming Dynasty coal mine, 1637, published with permission from Needham Research Institute; (b) coal waggons loading at a pithead, Tienfu (天府) coal mine, Jialing River (嘉陵江), 1944, published with permission from Needham Research Institute; and (c) China coal production, 1950–2020.*

Figure 4.3(c) gives an indication of the later boom in the Chinese coal industry under China's Gòngchǎndǎng government, from 1950 to the present day. This has proved to be the engine of the current expansion of China's economy.

4.2.1.4 North America oil developments

- ***Exploitation (see Figure 4.4)***

Oil exploitation became more active in the nineteenth century in many parts of the world but developed particularly rapidly in the northern states of the

USA and southern parts of the Canadian Great Lakes, for the following reasons:

- Known evidence oil seeps in the states of New York and Pennsylvania, Figure 3.8(a);
- Availability of relevant new scientific and engineering knowledge amongst local people, including Edwin Drake, to exploit these resources with extraction wells, Figure 3.8(b);
- The availability of significant capital funds, in the USA and United Kingdom, to invest in the exploitation of these oil resources;
- The potential exploitational opportunities for demand burning this oil for the cheap lighting of streets, private houses and industrial plants in the expanding cities of the northeast of the United States, particularly in New York and Philadelphia;
- This exploitation overlapped with the American Civil War. But the war did not disrupt technical and financial support for the activity, thus demonstrating an overwhelming demand in the North American economy at that time for safe and simple lighting in houses and factories.

This demand for oil products quickly expanded to include Great Britain, Ireland and Europe.

Modern coal, oil and gas explorations began in the nineteenth century and eventually came to depend upon highly sophisticated scientific technologies to detect and determine the extents of deposits, using explorational geophysics; see Figures 3.2 and 3.3.

This was the development of applied branches of geophysics and economic geology, using physical means at the Earth's surface, such as seismic, gravitational, magnetic, electrical and electromagnetic methods, to measure subsurface physical properties.

However, initial oil and gas exploration was highly competitive, incredibly destructive of the physical environment and limited by the technologies then available.

- ***Early fire risks, storage and networks***

In the early US oil industry, storage was based entirely on the use of wooden oil barrels, Figure 4.5(a).

Oil distribution networks in North America in the 1850s and 1860s were initially dependent on horse-drawn waggons loaded with these filled but leaky wooden barrels, driven by teamsters on un-tarred roadways to larger shipping points on rivers.

The teamsters had techniques for controlling the oil distribution trade, increasing their earnings, lowering the well-head value of the oil and raising its price as a fuel in the marketplace.

The lack of knowledge or well-pressure measurement leads to many early oil well blow-outs and the use of local barrel storage at well-heads leads to many oil-field fires and consequential accidents, Figure 4.5(b).

(a)

(b)

Figure 4.4 *Start of industrial oil exploration, Titusville, Oil Creek, Pennsylvania, USA. (a) Natural oil seep; (b) early chaotic Oil Creek well development, Edwin Drake c 1860s. Courtesy: Pennsylvania Historical and Museum Commission.*

However, the US oil industry innovated relatively quickly and this led to the laying of networks of small-diameter oil pipelines, typically steel pipes 2 inches in diameter, from wells to nearby rail-heads, which initially used rail-waggons to carry filled barrels of oil to the market.

In the second half of the nineteenth century, this led the US rail industry to act as carriers for oil barrels and then to introduce purpose-designed rail oil tankers. There was fierce competition between railroad companies to lay tracks close to oil production areas so that railroad companies could compete effectively in the lucrative oil transportation business.

(a)

(b)

Figure 4.5 *Hazards of early US oil exploration and transportation, 1860–92. (a) Early wooden oil transport barrels, 1860. Courtesy: Drake Well Museum and (b) oil-field fire at Titusville, 1892. Courtesy: Crawford County Historical Society.*

There was considerable disquiet in the USA of the increasing power of the railroad and oil industries due to their practice of forming financial 'trusts', which created problems for the national political control of oil development.

For example, Standard Oil consolidated its control over a wide range of disparate oil and rail businesses in an empire consolidated by trusts, a method that was anathema to Republican Senators who believed this ran counter to the popular values of United States' post-revolutionary republicanism.

- ***Consequences of early North American oil boom***

The North American oil boom was an exceptional development of very significant resources, causing a rush of ill-judged exploitation in relatively shallow depths of less than 100 m.

It led to not only the growth of a new industry but also the despoliation of pristine forest and riverine areas in north New York and Pennsylvania states:
'In the Oil Region itself, the invading multitude transformed the once quiet backwoods into a toxic, perilous wasteland where chaos ruled', Fisher [35].
The oil product was initially exploited primarily for the lighting market in the newer East Coast cities of the USA, with subsequent substantial exports to Europe, but the industry was beset by many damaging experiences and practices, including the following:

- A lack of planning, safety awareness or regulation in oil well exploitation;
- These led to unexpected fires, damage to property, pollution of rivers, streams and soil and disruption of adjacent oil sources. These were caused by excessive leakage due to badly regulated drilling, poor-quality oil flow management and inadequate fire and safety protection;
- The effects were aggravated by ill-planned exploitation, poor-quality well-drilling and inadequate transportation to bring the oil to the market, so serious that oil producers became at the mercy of teamster carriers, who exploited their monopoly by excessive charges;
- These effects consequently led in the early days to excessive swings in oil production and reduction in profits due to the disruptions.

Strangely the early oil industry expansion led to the laying of relatively few short, low-diameter pipelines, connecting well-heads to local rail-heads.

The railroads concentrated on the development of oil-carrying capacity, ignoring passenger traffic; the oil trade was initially too lucrative to leave much room for passengers!

But this growing oil industry also resulted in a substantial expansion of North American railroads.

This expansion in the USA was rather different from that in Great Britain and Ireland, probably because the newness of the USA meant that their cities had more newly settled populations. The huge demands for coal and oil transportation displaced railroads' attention from attracting travelling passengers, leaving long-distance passenger rail-travel in North America to be developed towards the end of the nineteenth century.

It was not until the late nineteenth century that the USA and Canada laid long-distance oil pipelines to transport the oil product to the remote river and sea-ports, which subsequently become the norm for oil production areas elsewhere in the world.

The great British railway developer, Sir Morton Peto, who in Britain built much of East Anglia's railways and is commemorated at Norwich station, wrote in 1865 from his experience in the USA about oil delivery:

'The great difficulty which has hitherto attended the oil trade has been transportation. It was easy enough to produce oil, but far more difficult to convey it to the consumer'.

Thus, the USA and Canada's early lead in world oil exploration and exploitation eventually came to help the development of a more closely regulated international environmental development of hydrocarbon resources as other oil-producing countries' production grew, subsequently outstripping the USA and Canada's production.

Modern methods of exploring the Earth's geological substructure in the twentieth century extended to much greater depths below the surface, revealing many more attractive locations where hydrocarbons are stored in the Earth's geological structure.

In the long term, oil and gas production profits steadily improved, despite periods of instability, becoming very considerable and laying the foundations for a strong North American oil industry resulting in greatly improved exploitation methods.

4.2.1.5 Russian oil developments

Oil was discovered in the Caucasus, in Baku and elsewhere, and Russian engineers exploited these resources with strong encouragement from sections of the imperial government. British engineers formed a significant part of the expertise needed to exploit these Russian wells.

Some of the problems of earlier North American exploitation had been learnt and mitigated, for example a stronger safety culture was beginning to emerge in Russian oil exploration. However, accidents could still happen; see Figure 4.6 in Baku in 1896.

Figure 4.6 Oil-field fire at Baku, during Russia exploitation, c. 1896

Standard Oil also tried to infiltrate this oil market but was rebuffed by Russian operators.

Russia's petroleum industry is now one of the largest in the world, it has the world's sixth largest oil reserves, and it is the world's largest exporter of natural gas and one of the largest oil producers.

Russia is currently the world's fourth largest energy user.

4.2.1.6 UK coal, oil and gas developments

- *Princess Victoria visits the Midlands*

In 1842 the UK's future Queen, Victoria, Figure 4.7, at the age of 13 recorded in her diary a visit by train to Birmingham and Wolverhampton, at that time the heart of the Industrial Revolution and now known as the Black Country.

Victoria wrote about her visit:

> *The men, women, children, country and houses are all black. But I cannot by any description give an idea of its strange and extraordinary appearance. The country is very desolate everywhere; there are coals about, and*

Figure 4.7 Princess Victoria, before her visit to the Black Country

the grass is quite blasted and black. I just now see an extraordinary building flaming with fire. The country continues black, engines flaming, coals in abundance, everywhere, smoking and burning coal heaps, intermingled with wretched huts and carts and little ragged children.

What Victoria was seeing form her train window was a consequence of the United Kingdom's burgeoning nineteenth-century Industrial Revolution, based upon coal, iron and gas, something like what can be seen in Figure 4.8(a) and (b).

- *Consequences of UK coal, iron, oil and gas developments*

Around 1800–14 an early ironworks in Coalbrookdale, England, was using coal to make iron and steel, Figure 4.8(a).

(a)

(b)

Figure 4.8 Coal, iron and gas developments, United Kingdom. (a) Iron, wrought iron and steel manufacture at Coalbrookdale, Staffordshire, United Kingdom, 1800; (b) UK Industrial Revolution, coal miner with background showing steam locomotive, coal waggons and steam-powered mine machinery, 1814.

The impression from the United Kingdom in the first half of the eighteenth century of burgeoning coal, iron and gas industrial developments is demonstrated in 1814 and shown in Figure 4.8(b), harking back to earlier mining activity in the sixteenth-century China (Figure 4.3(a)) or more recently in the twentieth century (Figure 4.3(b)).

These images all convey some of the features which the young Princess Victoria witnessed on her journey to Birmingham in 1842: smoking, flaming engines and apparatus, hard-pressed and bowed workers, coal heaps and pollution.

Interestingly the expansion of Great Britain's and Ireland's railway systems initially depended heavily on capturing these lucrative coal markets, moving the product from the pithead to cities, factories or ports on nearby rivers or the sea, via the railway.

This coal-trade gave an enormous stimulus, in the second half of the nineteenth century, to the United Kingdom's newly formed great railway operating companies like the London Northwestern, Great Northern and North Eastern Railways, enabling them to use coal-carriage revenue to generate capital to exploit and develop a rapidly expanding rail passenger service.

It was clear that, by the middle of the nineteenth century, there was a rising demand for long-distance passenger travel between long-established Great British and Irish cities, which the railway companies were eager to exploit.

This was quite different to American experience, where the expansion of oil traffic by rail in the 1860s did not lead onto improved passenger transport, which came 50 years later, as the US population expanded.

- *Exploitation of oil and gas*

The United Kingdom's exploitation of oil and gas followed a different trajectory to that in the USA but remained closely behind it, learning many valuable lessons from chaotic US blow-outs, fires and excessive well-head competition.

The United Kingdom's oil exploitation did not develop in its own country or waters; that occurred much later in the twentieth century, but in those other areas of the world where British overseas power had spread, particularly in Iraq and Persia, coming into competition with a growing Russian oil production industry in the Caucasus.

But UK companies also prospected for oil in Burma, present-day Myanmar, where oil traces had been found in the Irrawaddy River and surrounding areas, see Figure 4.9.

Demand for oil, in the 1870s–90s, from British sources was again initially for the supply of light oil, kerosene, for lamps in cities, in competition with American suppliers.

Standard Oil was also keen to enter this market and gain access to Southeast Asian oil sources.

The Netherlands was prospecting for oil in their South Asian territories, in Sumatra, Java and Borneo. The United Kingdom also invested in Dutch territory oil prospecting in Sumatra and Java, including collaboration with the Royal Dutch oil company, which later lead to the creation of the powerful Anglo-Dutch oil

Figure 4.9 Nineteenth-century British oil field in Burma, now Myanmar, showing worker ready to be lowered down a well with air supply

company, Shell, capable of competing with and resisting the United States' Standard Oil.

The United Kingdom's influence in oil was becoming less concerned with supplying lamps in UK cities, as in the USA, but with the fuelling and supporting of a worldwide fleet of UK merchant ships and a powerful and expanding Royal Navy, formerly dependent on coal-fired steam propulsion, supplied through a costly, complex network of coaling stations and numerous collier ships around the world.

The Royal Navy made its first oil fuel purchase for ship propulsion in 1899, from Shell's heavy oil wells in Dutch Borneo.

In the early part of the twentieth century, an increasingly important part in accelerating this naval fuel transition from coal to oil was played by Winston Churchill, who served in the British Government as President of the Board of Trade, Home Secretary and then First Lord of the Admiralty.

4.3 Outcomes from use of coal

4.3.1 General effect of coal's exploitation

Coal had been used as a more intensive fuel than wood for heating and the smelting of metallurgical ores for the manufacture of copper, bronze, brass, iron and steel, but it also contains other valuable minerals.

Samuel Smiles [3] reported both the deleterious and beneficial effects of these new developments in England's Black Country.

Coal became a mainstay for heating homes in Asian and northern European cities from the tenth century, and possibly earlier, replacing the use of wood because of coal's much higher calorific value.

Metals were initially being extracted from ores by smelting not only for the purposes of decoration but also for the manufacture of weapons and tools for agriculture and more recently in support of the means of land and sea transport.

Early coal extraction was from small, relatively local outcrops, at or close to the surface and near sites of human habitation or close to roads, rivers or the sea to permit transportation.

There is also some historic evidence of deeper mining down to 10–50 m, limited by human strength and the availability of ladders and lifting gear.

The following is a narrow view, based on the author's experienced of two widely differing countries, the British Isles and China, but gives a flavour of how humans have exploited valuable local sources of energy.

4.3.2 China

Ancient peoples were exploiting coal in China around 6,000 years ago; historians believe that by 3490 BCE they were involved in surface coal mining, pioneering the pre-modern world.

Natural gas was thought to have been first discovered in the Middle East between 6000 and 2000 BCE when lightning strikes ignited natural gas seeping from below the ground. However, in Beijing (北京) China, as early as 400 BCE workers are recorded as using crude bamboo pipelines to transport gas seeping from the Earth's surface to transmit it for burning to light the city.

Gas seepage also occurred close to China's coastline, where it was piped to shore to heat and boil seawater to obtain potable water.

From the sixteenth century, in Szechuan province, gas became increasingly closely associated with salt production. Gas was distributed through bamboo pipes and burnt to heat the salt pans for salt extraction from where salt wells release, Fisher [35], an echo of China's earlier gas usage.

Coal had started to be exploited industrially in China in the thirteenth to fifteenth century.

The first electricity generation source in China was for lighting the Summer Palace of Empress Dowager, Cixi, installed in Beijing by Siemens in 1900.

The first hydroelectric power plant in China was at Shilongba, Yunnan Province, with a capacity in 1912 of 240 kW.

China's first large industrial coal-fired power station was operational in Shanghai (上海) in 1920, with an output of 20 MW.

In the early twentieth century a similar sequence followed in China's coal industry as had occurred in eighteenth- to nineteenth-century Europe, with the consequential development of coke-making combined with steel manufacture for industry and a dependent coal-gas industry for lighting and other engineering applications.

By 2021 China's annual electricity demand has grown now to 8,310 TWh, with an installed plant capacity of 2,420 GW, including 1,118 operational coal-fired power plants.

But 45% of generation, 1,100 GW, is from non-fossil fuel sources, particularly hydroelectricity because of the accessibility of China's large rivers.

4.3.3 British Isles

Coal had begun to be used in iron-working industries in the British Isles after the eleventh century, being transported by carts and waggons to customers with an increasing river and sea trade becoming established, initially using small rowing boats in rivers and sailing vessels at sea around the coast as demand increased.

Coal mining activity was concentrated around Newcastle and South Wales, although smaller outcrops may have been exploited locally in other coal-rich areas. Eventually, by the fourteenth century, a substantial coal sea trade became established from Newcastle to London, increasing in subsequent centuries and becoming industrialised as purpose-built coal-bearing sailing vessels, known as colliers, were introduced.

By the seventeenth century, a substantial fleet of colliers were working around the coasts of the British Isles, supplying coal to London, Edinburgh, Glasgow and Dublin from the mining centres in the NE England and South Wales.

The English naturalist and clergyman, Gilbert White, recorded in his book 'The Natural History of Selbourne' [1] the effects on his local UK Hampshire environment of airborne pollution, caused by coal burning in the capital.

By 1795, 4,395 coal cargoes were being delivered by colliers to London per year, by 1824 this number had risen to about 7,000 and by 1839 to over 9,000 per annum.

In 1865 an interesting observation was made by W S Jevons in his book '*The Coal Question*' [4], where he estimated that, at the then current rate of UK coal extraction, by 1965 Britain would be taking more coal than lay underground in the country. Ultimately this situation did not occur in United Kingdom, where pressures from European governments and anti-pollution groups forced the UK Government to reduce fossil fuel firing of power stations.

It is interesting, however, to speculate at what has happened elsewhere, in much larger countries, such as India or China, where industrialisation continues to increase, outstripping indigenous fossil fuel resources, forcing those countries to import fossil fuels from abroad, largely from Russia, Australia and South America.

In the first half of the seventeenth century in the British Isles, gas became an important fuel, as a by-product of the coal industry, where carbon from coal was beginning to be used for the expanding steel industries. The carbon was extracted by coking high-performance coal, and a natural by-product of coke manufacture was a stream of coal-gas, which became available as a fuel, initially for lighting but later for firing gas engines.

The exploitation of coal in Great Britain and Ireland expanded enormously during the eighteenth and nineteenth centuries, going deeper underground via pithead vertical shafts with workings that advanced into the coal seams a long way from those vertical shafts. The coal seams were of varying depths, slopes and heights, which ranged from 1 to 6 m, and up until the nineteenth century, they were entirely exploited by the use of hand tools, with extraction to the pithead vertical shaft via man-hauled waggons running on simple wooden tracks.

However, in the nineteenth century as coal's value increased, technology advanced, with the use of steam power to lift the coal at the pit-shaft and the use of ponies and improved waggons running on iron and then steel tracks underground.

Safety risks increased as deeper seams were exploited operating a long way, sometimes many kilometers, from the vertical pithead shaft. However, as risks increased, safety measures improved in the twentieth century, drastically reducing fatalities. These were achieved by improvements to miners' equipment in terms of helmets, the use of safe miners' lamps, the more systematic measurement of fire-damp or methane and the development of skilled mine-rescue teams.

Before coming to the next section, about the development of sciences, there was an amusing vignette, in the early nineteenth century, about scientific development, concerning the issue of underground illumination in coal mines.

By then coal miners worldwide used small wick lamps to guide their ways through coal measure excavations. But emissions from these measures, particularly methane, made this hazardous as a naked lamp flame could ignite that gas, initiating a dangerous fire or underground explosion.

Both the fashionably eminent scientist, Humphrey Davy, and the bluff engineer, George Stephenson, independently developed, in 1815, miners' safety lamps for use in coal mines to prevent methane ignitions, Figure 4.10.

Stephenson's lamp, affectionately known by miners as the 'Geordie', was widely used in the Great Northern coal field in County Durham and Northumberland, United Kingdom, adopting an original design, significantly different from Davy's patented lamp, which used a close-mesh gauze barrier to isolate the lamp flame from external methane, using a suitable chimney. Both lamps were sealed, except for restricted air ingress, so dangerous amounts of methane in the incoming air would, by combustion, reduce oxygen concentration inside the lamp, extinguishing the flame.

Davy's prestige was such that his lamp was sold worldwide, and as a rather privileged intellectual, he was furious with Stephenson for the 'Geordie', the product of another gifted but practical man, believing him to have breached his patent.

But history has shown that, surprisingly, both these two diversely trained men had reached the same conclusion, at the same time, but by different methods!

However, the enormous size of the British coal industry meant that, despite using both Davy and Geordie lamp technology, there were still occasionally catastrophic underground explosions.

The worst UK incident was an explosion on 14 October 1913 at the Senghenydd colliery near Caerphilly, Glamorgan, Wales, killing 439 miners and 1 rescuer.

Many of the twentieth-century improvements in safety in British coal mines were achieved by close collaboration between mine owners and miners trade

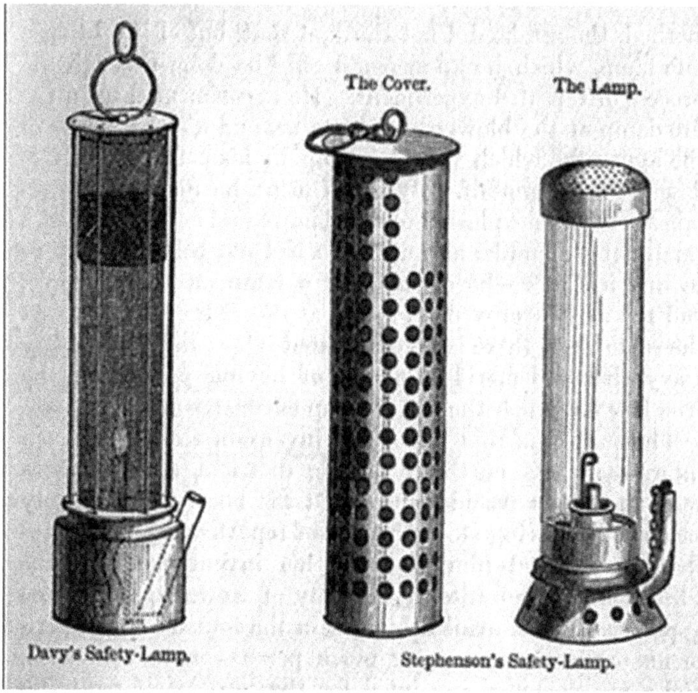

Figure 4.10 Coal miners' safety lamps, nineteenth century

unions; that was particularly so in the UK after the nationalisation of the coal industry following the Second World War.

In the second half of the twentieth-century British coal extraction slowed for the following reasons:

- Reducing exploitable coal measures;
- Cheapness of coal available from international sources, for example Southeast Asia, Africa, South America and Russia;
- Growing concerns about the environmental effect of pollution from UK coal-fired power stations, not only in the United Kingdom but also because of complaints from European neighbours, about that air pollution, aggravated at that time during UK's membership of the European Community;
- The thermodynamic efficiencies of oil and combined cycle gas-burning plants were gradually trumping those of coal-fired plants because of their superior thermodynamic performance;
- Oil and gas were proving to be cleaner, more abundant and easily managed fuels than coal, Fisher [35].

As a CEGB engineer, the author witnessed two interesting UK air pollution vignettes, in the 1980s, while providing scientific services to large coal- and oil-fired power stations along the Thames estuary.

Tilbury B was a Thameside power station with two chimneys and four boilers, burning poor-quality East Midlands coal, supplying electricity to the Grid from 4×350 MW generator units. The power station's efficiency record was not high, about 27%–28%, placing it low in the merit order at that time, so the station was not permitted to operate continuously. But in 1981 the station fought to remain operational, becoming skilful at daily two-shifting, accepting load early in the morning from 0530 to 1100, then taking units offload and accepting load again from 1800 to 2300. This was organisationally and technically a complex process but achieved on a daily basis. However, grid control was critical that Tilbury's two-chimney stack emissions could be high because of their use of low-quality coal. The author advised the station that by using their good-quality stack emission monitoring and sound boiler firing control, it was possible for the station to demonstrate directly to grid control, from their emission readings, that they were demonstrably within the limits demanded, which was at that time less than 10 g SO_2/kWh.

A later, related experience occurred during the 1984–85 UK miners' strike, for which the Government had ordered the CEGB to stock-pile coal. Once that coal was depleted during the winter, grid control relied for London electricity supply on the two large Thames-side, high efficiency, over 33%, oil-fired power stations Littlebrook D, 1,980 MW, and Grain, 1,320 MW, in preference to relying on the depleted coal-fired Tilbury B.

Littlebrook D and Grain power stations each had modern, high, concrete chimneys. However, when cold, low-pressure weather conditions prevailed, during the winter of 1984–85, with wind persistently in the East, these power stations' smoke plumes coalesced over London, causing SO_2 readings in the city to rise, occasionally to exceptionally high figures of more than 20–25 g SO_2/kWh.

The UK coal industry came to a virtual end by 2005.

By that time, the huge Drax coal-fired power station, 3,300 MW, in Yorkshire, UK first operational in 1974, was being supplied with coal from five nearby mines, using a sophisticated system of underground and pithead surface conveyor belts with merry-go-round trains operating between those pitheads and the power station.

Those five pits were exhausted within 25 years, less than two-thirds into the power station's prospective life.

However, the power station's huge plant investment meant that it still needed to continue operating, while changing fuel and cleaning up its exhaust emissions, as far as possible, by making use of biomass and other cleaner fuels.

Drax power station continues operating today, with four boiler units burning woodchip, imported from the United States and elsewhere, mitigating its carbon footprint by extracting CO_2 from the boiler exhaust stream and two boiler units continue to be capable of burning imported coal, when needed.

There are also other mothballed UK fossil-fired power stations, still available for occasional duties:

- In England, West Burton and Ratcliffe-on-Soar on the river Trent and its tributary;
- In Ireland, Kilroot on the Belfast Lough.

These power stations are available to operate, burning coal or oil, typically in the winter when electricity demands are high and supplies short. Fuel for these power stations is being imported from Australia, Colombia, Poland or South Africa and oil is imported from the North Sea or Middle East.

In the nineteenth century, the United Kingdom had identified within its borders hard coal resources of 3,910 million tonnes, although total resources could be as large as 187 billion tonnes. There were 33 million tonnes of economically recoverable reserves available at operational and permitted mines, plus a further 344 million tonnes at mines in planning.

Whilst the majority of this accessible UK coal has now been extracted, one of its largest resources, the Great Northern coal field in Northumberland and County Durham, extends under the North Sea, containing high-quality, hard bituminous coal, difficult to extract, but 75% of its former total capacity still remains unmined.

Perhaps this resource could be exploited in future times as a feedstock for chemical industries rather than being employed as a raw fuel in power stations with thermal efficiencies lower than 38%!

By 2021 UK's annual electricity demand had grown to 334 TWh, with a total installed plant capacity of 76 GW. Fossil fuel generation capacities, from Drax or CCGT plants, and nuclear capacity has remained largely unchanged from previous years at 43 and 8 GW, respectively, while renewable capacity had increased to 23 GW.

4.4 Scientific rationalisation of energy from fuels

The following sections describe how in the seventeenth, eighteenth and nineteenth centuries a scientific revolution was gradually taking place, leading to very significant scientific rationalisations, largely arising from the increasing fuel usage described earlier. These rationalisations could ultimately reverse the deleterious effects of industrialisation and inefficient use of these fuels. The following sections describe these scientific changes and those who brought them about.

4.4.1 Mechanics

Heron of Alexandria was a Greek mathematician and engineer, c. 60 AD, active in his native city of Alexandria in Egypt during the Roman era, Figure 4.11(a). He developed a simple steam turbine, as a plaything, describing his construction as Heron's engine or aeolipile, Figure 4.11(b), a reaction turbine, nearly two millennia before the Industrial Revolution.

In Italy, Galileo Galilei, Figure 4.11(c), had started in the sixteenth century the rationalisation of modern science. He studied speed, velocity, gravity and free fall, the principles of relativity, inertia and projectile motion. He also worked in applied science and technology, describing the properties of

(a)

(b)

(c)

(d)

Figure 4.11 The dynamicists. (a) Heron, Greek Scientist, first century AD; (b) Heron's aeolipile, an early reaction steam turbine; (c) Galileo Galilei, 1564–1642, Italian astronomer, physicist and engineer, sometimes described as a polymath; and (d) Isaac Newton, 1642–1726, English mathematician, described natural philosopher, painted by Godfrey Kneller.

pendulums and what he called hydrostatic balance. He invented the thermoscope, various military compasses and used the telescope for the scientific observation of celestial objects. His contributions to observational astronomy include telescopic confirmation of the phases of Venus, observation of Jupiter's four largest satellites, observation of Saturn's rings and the analysis of lunar craters and sunspots.

Meanwhile in England the mathematician Isaac Newton, Figure 4.11(d), developed the equations of motion for mechanics at Cambridge University, marking the start of his rise to become a towering late seventeenth-century scientific figure:

- A body remains at rest, or in motion at a constant speed in a straight line, unless acted upon by a force;
- When a body is acted upon by a force, the time rate of change of its momentum equals that force;

- If two bodies exert forces on each other, these forces have the same magnitude but opposite directions.

He gradually unfurled a mathematical representation predicting the dynamics of the planets and for objects on Earth, see Gleich [17].

Newton was influenced by Galileo's work and other British and European scientists, including Halley, Hooke and Leibnitz.

He was secretive but sound, with a gradually developing grasp of the fundamentals of the mathematics and dynamics of the world and our universe, translating his ideas into modern mathematical expressions, which he slowly displayed to the eighteenth-century European scientific world.

In 1696 Newton went on from Cambridge to become Warden of the Royal Mint, introducing a new coinage for England and restoring the probity of the nation's currency.

He became in England the President of the Royal Society from 1703 to 1727.

His magnum opus, 'Philosophiae Naturalis Principia Mathematica', was printed under the seal of the Royal Society, and has been reproduced, in several issues of increasing clarity.

4.4.2 Mathematics

At the start of the nineteenth century, the Cambridge mathematician, philosopher and inventor Charles Babbage, 1791–1871, who was an astronomer, originated the concept of a digital programmable computer, developing a ten-digit mechanical difference engine to assist in his astronomical calculations. He was advised on his methodology by Ada Lovelace, 1815–52, the pioneering daughter of Lord Byron and mathematician; see Figure 4.12.

Babbage also planned a larger, more complex analytical engine, which was a ten-digit mechanical forerunner of the modern binary electronic computer.

4.4.3 Hydrodynamics

Archimedes, Figure 4.13(a), a Greek mathematician, physicist, engineer, astronomer, and inventor is originally from the city of Syracuse in Sicily. A few details of his life are known but he is regarded as one of the leading scientists of classical antiquity, initiating our understanding of water.

William Froude, Figure 4.13(b), a nineteenth-century British engineer developed our understanding of the effect of water on floating vessels. He built the first tow-tanks in which ship models could be tested to improve their hull shapes and enhance their performance in varying wave conditions. Towards the latter half of the nineteenth century, he was an adviser to the Royal Navy and contributed directly to the development of the preferred high-speed warship hull shapes.

4.4.4 Chemistry

In chemistry, key players were Robert Boyle, Joseph Priestley, Henry Cavendish and Humphrey Davy and Marie Skłodowska-Curie; see Figure 4.14.

Figure 4.12 Early computing, Charles Babbage and Ada Lovelace. (a) Charles Babbage, mathematician and inventor; (b) Ada Lovelace, mathematician; (c) Babbage's difference engine, 1832; (d) Babbage's analytical engine, 1834–71.

Robert Boyle was an Anglo-Irish natural philosopher, chemist, physicist, alchemist and inventor.

Boyle is regarded today as the first modern chemist, and therefore one of the founders of present-day chemistry and pioneer of modern experimental scientific methods. He is best known for Boyle's law describing the inversely proportional relationship within a closed system between gas volume and absolute pressure when temperature is kept constant.

Among his works, *The Sceptical Chymist* is seen as a cornerstone in the field of chemistry. He was a devout and pious Anglican also noted for his writings on theology.

Joseph Priestley was an English chemist, natural philosopher, separatist theologian, grammarian, multi-subject educator and liberal political theorist.

(a)

(b)

Figure 4.13 *The hydrodynamicists. (a) Archimedes of Syracuse, 287–212 BC, Greek mathematician, physicist, engineer and inventor; and (b) William Froude, 1810–79, English engineer, hydrodynamicist and naval architect.*

He published over 150 works and conducted experiments in electricity and other areas of science.

He was a close friend of Benjamin Franklin and worked in association with him on electrical experiments. Priestley is credited with his independent discovery of oxygen by the thermal decomposition of mercuric oxide, having isolated it in 1774. During his lifetime, Priestley's considerable scientific reputation rested on his invention of carbonated water, his writings on electricity, and his discovery of several gases, the most famous being oxygen, which Priestley dubbed 'dephlogisticated air'.

Priestley's determination to defend the phlogiston theory, rejecting what became the chemical revolution, eventually left him isolated within the scientific community.

Henry Cavendish was an important preliminary figure in eighteenth-century English science, who combined a knowledge of chemistry with an interest in electricity.

He pioneered work on electricity at Cambridge.

He was also one of the founders of the Royal Institution in London in 1799.

Later in the nineteenth century, after James Clerk Maxwell had been recruited to the Cavendish Chair at the university, he took great trouble to document Cavendish's work and express his appreciation of its value.

Humphrey Davy was an important socialite in London, a scientist and an encourager of other engineers and scientists in the nineteenth century. He expanded

Figure 4.14 *The Chemists. (a) Robert Boyle, 1627–91, Anglo-Irish natural philosopher, chemist, physicist and inventor. (b) Joseph Priestley, 1733–1804, English chemist, natural philosopher and separatist theologian. (c) Henry Cavendish, 1731–1810, English natural philosopher, scientist experimental, theoretical chemist and physicist, inspiration to Maxwell. (d) Humphrey Davy, 1778–1829, British chemist and inventor, responsible for the Davy and early arc lamps, Michael Faraday's encourager. (e) Marie Skłodowska-Curie, 1867–1934, Polish/French chemist and inventor, researcher on radiation phenomena.*

the work in London of the Royal Institution as a place where young people could learn about science and engineering.

Davy was primarily a chemist and made many important scientific interventions, including the development of the coal miners' safety lamp described earlier, Figure 4.9.

James Watt knew Davy and designed and built a gas chamber to facilitate Humphrey Davy's experiments for the inhalation of nitrous oxide.

Perhaps Davy's most important contribution to science was to recruit and encourage the young Michael Faraday [2], who started work as Davy's assistant. He became a famous lecturer at the Royal Institution and initiated a profound understanding of electromagnetism, pioneering the work of Maxwell [5] and Heaviside.

Marie Skłodowska a Polish schoolgirl attending the J. Sikorska boarding school and gymnasium for girls in Warsaw, graduated in 1883 with a gold medal. She then studied and tutored at the clandestine Flying or Floating University, a patriotic Polish institution of higher learning in Warsaw that admitted women. In 1891 she came to Paris and began her scientific career, investigating the magnetic properties of various steels. That same year Marie Skłodowska and Pierre Curie met, and through their mutual interest in natural sciences, they were drawn together and married. Curie decided to investigate uranium rays as a possible field for thesis research, using an innovative technique to investigate samples. Her husband and his brother had developed an electrometer 15 years earlier, a sensitive device for measuring electric charge. Marie Skłodowska-Curie observed, using her husband's electrometer, that uranium rays caused the air around a uranium sample to conduct electricity, discovering the ionising effect of radiation.

4.4.5 Thermodynamics and heat engines

In 1824, Nicolas Léonard Sadi Carnot published a short book devoted to the study of heat engines, *Réflexions sur la Puissance Motrice du Feu et sur les Machines propres a développer celle Puissance*, read at a meeting of the French Académie Royale des Sciences, attended by the best of contemporary French Science, Arago, Fourier, Laplace, Ampère, Gay-Lussac, Poinsot, Fresnel, Legendre, Poisson, Cauchy, Dulong, Navier and Riche De Prony. However, that book had to wait 10 years before it was read by a practising engineer, Emile Clapeyron, and another 10 years elapsed before the Scots/Irish scientist, William Thomson, came across Clapeyron's treatment of Carnot's ideas.

Klein (1974) has shown how influential Carnot's ideas were to Thomson and how Clausius, motivated by Thomson's endless questioning, laid the foundations of modern thermodynamic theory, as a synthesis of the law of conservation of energy with principles first formulated by Carnot.

However, it is no less true that Sadi Carnot 'conceived the very categories of thermodynamic reasoning'. Among them, Carnot's principle states that the operation of heat engines consists of a transportation of heat from a hot source, such as a steam engine boiler, to a cold source, a steam engine condenser, and not by the consumption of heat.

Figure 4.15 *The Thermodynamicists. (a) Nicolas Léonard Sadi Carnot, 1796–1832, Founder of thermodynamics; (b) Benoît Paul Émile Clapeyron, 1799–1864, French engineer and physicist; (c) William Thomson, 1890–95, British, Irish mathematician, physicist and engineer. (d) Rudolf Clausius, 1822–88 German physicist and mathematician.*

Carnot, a scientist, developed the concept of the heat engine years before the development by engineers of coal, oil and gas-fired machines in various European and American countries.

The initiation in France by Carnot of the science of thermodynamics has had a profound effect on our development of and improvement of our fossil fuel-fired economies.

Another important discovery, related to Carnot's debt to contemporary technology, was made by Cardwell [12]. At that time, in the late eighteenth and early nineteenth centuries, the efficiency of hydraulic devices was measured by their capability to restore initial conditions.

For example, an engine was thought to have an efficiency of 70% if the power it generated could take 70% of the water used to operate the engine back to the reservoir, whenever the engine was used as a water pump.

Cardwell [12] concluded that this practice showed that the fundamental concept was the recoverability of the initial conditions and not the concept of reversibility.

The thermodynamic principles of Carnot, Clapeyron and Clausius enabled many of the key future energy exploitation developments in historical sequence, to be described in following chapters, including the following:

- Slow-speed reciprocating steam engines for pumping water out of mines;
- Higher speed reciprocating steam engines for locomotives;
- Slow-speed reciprocating steam engines for marine propulsion and for driving electricity generators;
- High-speed steam turbines for driving electricity generators;
- High-speed steam turbines for marine propulsion;
- High-speed gas or diesel engines for pumping and electricity generation;
- High-speed in-line petrol and diesel engines for driving cars and heavy lorries;
- High-speed rotary and then in-line petrol or diesel engines for aviation;
- High-speed gas turbines for aviation, which were then applied to electricity generation;
- Development of combined cycle gas and steam turbine arrangements for more efficient generation of electricity.

Persons referred to in this section are shown in Figure 4.15.

4.4.6 Magnetism, electromagnetism and electricity

The science of magnetism was developing in the sixteenth and seventeenth centuries, but the eighteenth-century electromagnetism and the generation of electricity were pioneered, as follows, Figure 4.16.

In Italy by Luigi Galvani, an Italian physiologist and philosopher, studying animal electricity in 1780, he described how frogs' legs, suspended from a metal lattice by metal hooks, twitched convulsively if the hooks were moved and the legs touched the lattice, postulating the presence of 'animal electricity'.

Galvani was followed by Alessandro Volta, an Italian physicist and chemist, discoverer of methane and pioneer of electricity. He was a Professor of Experimental Physics at the University of Pavia and among the first scientists to repeat Galvani's experiments, giving an improved scientific explanation. At first Volta embraced Galvani's concept of 'animal electricity'; however, he started to doubt that conduction was caused by electricity intrinsic in the frog's legs or body parts. He believed the frog's legs' contractions depended on the fluids present and the differing metals of the hooks and lattice used by Galvani to connect the frogs' legs' nerves and muscles in these experiments.

Based upon this development, Volta produced, in 1799, a voltaic pile using differing metals separated by cloth or card soaked in brine.

He reported his results, in 1800, in a two-part letter to the President of the Royal Society, Sir Joseph Banks.

(a) (b) (c)

Figure 4.16 The Electricians. (a) Luigi Galvani, 1737–98, Italian physiologist and philosopher; (b) Alessandro Volta, 1745–1827, Italian physicist and chemist who expanded Galvani's work, pioneering electricity; (c) Michael Faraday, 1797–1861, English scientist contributing to electromagnetism and electrochemistry.

In UK, Michael Faraday, 1791–1867, a self-taught scientist working as an assistant to Sir Humphry Davy, the British chemist and inventor of the miners' safety lamp and early form of arc lamp, later President of the Royal Society.

Because of the Davy's experience of Faraday's work, he was recommended to be Chemical Assistant to the Royal Institution, a society established to enhance scientific education and research, based in London in the City of Westminster, where Sir Humphry previously gave scientific lectures to young people interested in science.

Faraday had a natural gift for explaining complex principles to interested technicians; he eschewed the use of mathematics, employing clear explanations

(a) (b)

Figure 4.17 The rationalisation of electricity. (a) James Clerk Maxwell, 1831–79, Scottish mathematician, developer of electromagnetic theory; (b) Oliver Heaviside, 1850–1925, English rationaliser of practical electromagnetic theory.

in his lectures and books. His pioneering practical work on practical electromagnetism is set out in two parts, 'Experimental Researches in Electricity' [2], clear in exposition but without mathematical formulation, it is still worthy of study.

The mathematical groundwork for electromagnetism was then thoroughly codified, mathematically, by James Clerk Maxwell (1831–79), Figure 4.17(a), during his time at Aberdeen and Cambridge Universities then brought up to date in two parts 'A Treatise on Electricity and Magnetism' [5].

Maxwell was a poor lecturer but a fine, amusing and clear mathematical thinker; his works were a huge encouragement to scientists and physicists. For some reason, he had taken great trouble to document the work of his professorial predecessor at Cambridge, the chemist Henry Cavendish. However, eventually he settled down to his indispensable work on electromagnetism.

Maxwell's electromagnetic formulation in 1864 is a classic as follows:

- Gauss' law for electrostatics:

$$e + \frac{df}{dx} + \frac{dg}{dy} + \frac{dh}{dz} = 0$$

- Gauss' law equivalent for magnetism:

$$\mu\alpha = \frac{dH}{dy} - \frac{dG}{dz}$$

$$\mu\beta = \frac{dF}{dz} - \frac{dH}{dx}$$

$$\mu\gamma = \frac{dG}{dx} - \frac{dF}{dy}$$

- Faraday's law, with Lorentz force and Poisson's law:

$$P = \mu\left(\gamma\frac{dy}{dt} - \beta\frac{dz}{dt}\right) - \frac{df}{dt} - \frac{d\Psi}{dx}$$

$$Q = \mu\left(\alpha\frac{dz}{dt} - \gamma\frac{dx}{dt}\right) - \frac{dG}{dt} - \frac{d\Psi}{dy}$$

$$R = \mu\left(\beta\frac{dx}{dt} - \alpha\frac{dy}{dt}\right) - \frac{dH}{dt} - \frac{d\Psi}{dz}$$

- Ampére–Maxwell's law:

$$\frac{d\gamma}{dy} - \frac{d\beta}{dz} = 4\pi p' p' = p + \frac{df}{dt}$$

$$\frac{d\alpha}{dz} - \frac{d\gamma}{dx} = 4\pi q' q' = q + \frac{dg}{dt}$$

$$\frac{d\beta}{dx} - \frac{d\alpha}{dy} = 4\pi r' r' = r + \frac{dh}{dt}$$

- Ohm's law:

$$P = -\xi p Q = -\xi q R = -\xi r$$

- Electric elasticity equation:

$$P = kf Q = kg R = kh$$

- Continuity of charge:

$$\frac{de}{dt} + \frac{dp}{dx} + \frac{dq}{dy} + \frac{dr}{dz} = 0$$

However, Maxwell's mathematical approach was not necessarily easily accessible for working technical people, particularly an emerging breed of electrical engineers and technicians, as it included 20 electromagnetic equations, with many main variables and various subsidiary constants and his work 'A Treatise on Electricity and Magnetism' is still a tough read.

This gave an opportunity for a clever technician, Oliver Heaviside (1850–1925), Figure 4.17(b), a practising telegraphist, who operated some of the first telegraphic links between the United Kingdom and USA.

Heaviside [6] was a thinker, influenced perhaps by Faraday's earlier clear physical descriptions of electrical and magnetic processes and by his own practical experience. Heaviside's magnum opus, based upon Maxwell's principles, was set out in three parts 'Electromagnetic Theory' the practicalities of electromagnetism, enabling engineers and technicians to develop what became the network, communication and electrical machine industries of Europe and the USA.

Heaviside's 1893 formulation of electromagnetism, in vector form, is as follows, being preceded by Maxwell's 20 equation formulation. Heaviside's formulation has four equations:

- Gauss' law for electrostatics: $\nabla.\boldsymbol{D} = \rho$
- Gauss' law for electromagnetism: $\nabla.\boldsymbol{B} = 0$
- Faraday's law: $\nabla \times \boldsymbol{E} = -\partial \boldsymbol{B}/\partial t$
- Ampére–Maxwell's law: $\nabla\times\boldsymbol{H}=-\partial \boldsymbol{D}/\partial t+\boldsymbol{J}$

With five main variables:

- $\boldsymbol{D}$ is the electric displacement vector;
- $\boldsymbol{B}$ is the magnetic field flux density vector;
- $\boldsymbol{E}$ is the electric field vector;
- $\boldsymbol{H}$ is the magnetic field vector;
- ρ is the electric charge density.

This work earned Heaviside his Fellowship of the Institution of Electrical Engineers and subsequent membership of the Royal Society and has become a backbone for electrical engineering throughout the world.

His 'Electromagnetic Theory' is still a good read, but is peppered with acerbic flings at his opponents, which rather detract from the clarity and importance of this work!

The present relativistic formulation for the electromagnetic equations is as follows:

$$\partial_{\propto}\boldsymbol{F}^{\propto\beta} = \mu_0 \boldsymbol{J}^{\beta}$$

This formulation of one equation and two covariant four-vectors is as follows:

- $\boldsymbol{F}$ the electromagnetic four-potential, a covariant four-vector for charge, containing the electric scalar potential ϕ and magnetic vector potential $\boldsymbol{A}$;

- $\boldsymbol{J}$ the electromagnetic four-current, a contravariant four-vector for current densities, combining electric charge density ρ and electric current density $\boldsymbol{j}$;
- μ_0 is a physical constant;
- $\propto$ and $\boldsymbol{\beta}$ are suffixes.

4.4.7 Women in energy science and engineering

The author has added this section to identify an important twentieth-century scientific development, not directly connected to human energy usage, but because it was the major biochemistry development, demonstrating how the cells in living creatures managed their energy usage.

That event was the discovery and identification of the structure of DNA molecules, a complex international science undertaking involving two Nobel Prize winners but also an important contributor and equal player in the DNA science discovery process.

That contribution came in 1951 from the scientist and crystallographer, Rosalind Franklin, whilst working at Kings College, London, see Figure 4.18(a).

The author's point here is to highlight the increasing contribution of women scientists and engineers, in particular the work of a Durham colleague, Prof Karen Johnson, Figure 4.18(b). Her work on the engineering and rebuilding of soils has been awarded, in 2023, the Rosalind Franklin Prize from the Royal Society, emphasising the importance of soil recovery [36] with graphics, Figure 4.19, from that reference.

(a) (b)

Figure 4.18 Women in science and engineering. (a) Rosalind Franklin, British chemist and DNA molecular structure investigator; and (b) Professor Karen Johnson, investigating engineering soil strength at Durham University.

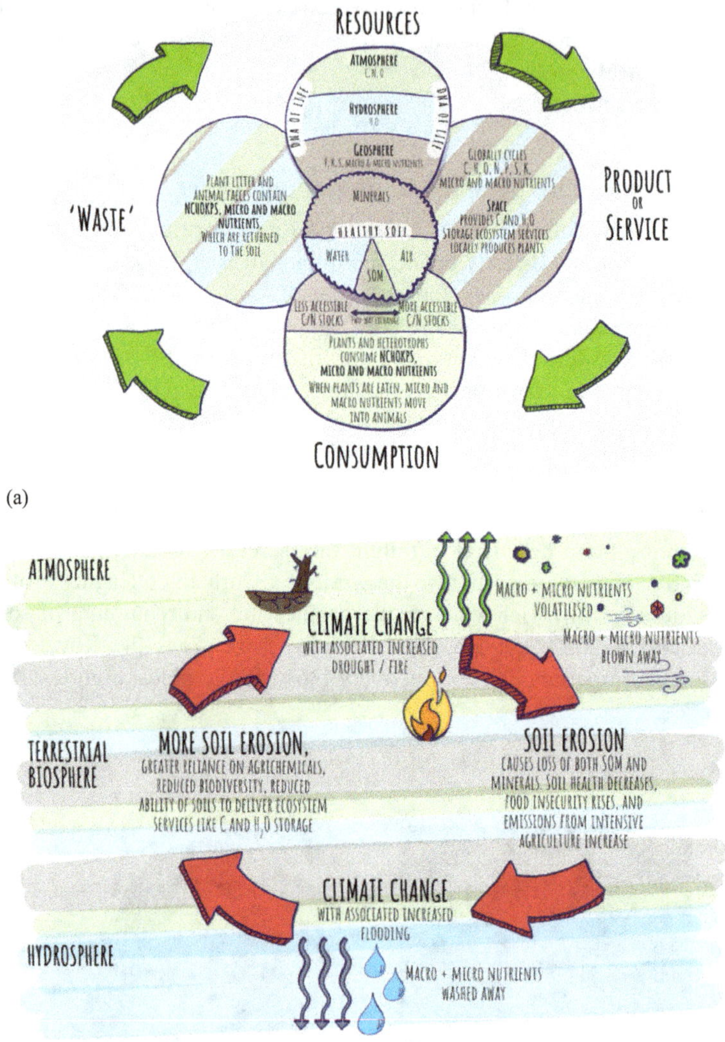

Figure 4.19 Rebuilding soils rebuilds a nation, an engineer's perspective. (a) Soil's natural cycle, shown as a 'circular economy' with resources, product/service, consumption and waste; and (b) vicious cycle of soil degradation and climate change.

4.5 Summary

This chapter has demonstrated that following early practical Industrial Revolution applications, fundamental scientific and engineering principles began being laid down, forming the foundations for modern science and the prospect of more efficient energy technology developments.

Their exploitation will be described in the following chapters in these key areas:

- Mechanics:
  Knowledge coming from the ancients began to evolve in the eighteenth century into rigorous science as greater demands were being made on our mechanical knowledge. The importance of mathematics and experimentation lead to rapid practical advances;
- Hydrodynamics:
  The needs of a growing ship-building and naval industry led to a greater understanding of ship's hulls shape and also of the best suited propulsion to enable worldwide steam ship operations.
- Chemistry:
  The science of chemistry grew because of the need for industrial chemicals. The English chemist Joseph Priestley made a major contribution. But the changes taking place in the science of chemistry, from older ideas, such a phlogiston, damaged the reputation of some of our pioneers. Marie Skłodowska-Curie also demonstrated the potential of radioactive materials and chemistry's connection to electricity.
- Thermodynamics, heat engines and energy:
  The pioneering eighteenth-century thermodynamics of Carnot and Clapeyron led to the nineteenth-century understanding of Clausius and Thompson and the development in the nineteenth century of efficient heat engines by Richard Trevithick, James Watt, George and Robert Stephenson;
- Electricity, magnetism and electromagnetism:
  The extraordinary nineteenth-century physical and mathematical understanding of electricity and electromagnetism by Faraday, Maxwell, Heaviside and many others has enabled the advancement of electrical generation, transmission and radio communication in the twentieth and twenty-first centuries;
- Rebuilding our soils can rebuild our nations:
  Damage to the Earth's soils has also been caused by climate change, but we can build and rebuild degraded soils to address that and potentially improve the nutritional content of our food if we change policies that allow the addition of safe mineral and organic wastes to soil.

Chapter 5

Modern age I, 1800–1930 CE, science, engineering and energy

5.1 Introduction

This chapter considers the effects of modern scientific knowledge from around the world on the conversion of energy from water, hydrocarbons and gases, as described in the previous chapter, and the importance of the scientific knowledge for the improved performance of machinery, particular its efficiency.

This will also cover the increasing effectiveness and transportability of fossil fuels by land and sea, but their deleterious polluting effects and the influence of energy efficiency on their economy and pollution.

The chapter will demonstrate the importance of heat storage and flow in energy deployment and of associated transport networks, including electricity, for sharing power and enabling its distribution from diverse energy sources. The development of renewable technologies to meet energy needs, without the need to use deleterious fossil fuels and the essentially distributive nature of these new energy resources.

The chapter will also show how modern communication technologies can also facilitate energy flows for improved manufacturing, planning, economic efficiency and development. Improved manufacturing techniques also mean that we can construct engineering facilities at lower cost, in shorter times and with less environmental impact than hitherto.

The surprising abilities of these new technologies mitigate some of the deleterious polluting effects of traditional energy sources.

The important part is played by electricity, oil and gas networks in the distribution and effective application of energy sources and particularly of the way that is likely to affect the future development of renewable technologies and of energy supply.

The possible influence of future Energy Vectors alongside electricity is made by hydrogen or other alternatives.

5.2 Energy vectors

5.2.1 Water

Water has been a source for food by fishing and hunting, for more than 10,000 years and a vector for transporting humans, children, animals and their food, so that human activity has spread from lakes to rivers and then to seas, to explore other continents.

The steady development of vessels to operate, initially on lakes but then on rivers and seas, has been a vital part of human technological development, see [13], and a basis for modern human development.

The great lakes on Earth are sources of food and the technologies needed to manage boats and larger vessels for trade.

Then trade, ideas, politics and technology have spread along the great rivers.

In China along the Yellow River (黄河) to the Bo Hai Sea and along the Yangtze (長江) down to the Dong Hai Sea.

From Scandinavia into Russia along the Volga and the Don rivers.

From North India along the Indus into the Arabian Sea and along the Brahmaputra and Ganges into the Bay of Bengal.

In Europe along the Rhine draining into the North Sea, the Danube draining into the Black Sea or the Loire draining into the Bay of Biscay.

Water can become an Energy Vector if it is heated to raise steam.

Or the waters of a river can be dammed to power a hydro-electric installation, or if we can develop the technology to harness waves, driven by the winds, or the tides, driven by the Moon and the Sun.

5.2.1.1 Lakes

The following are the World's largest lakes by geological definition, down to 17,500 km^2 by plan area:

In North America: the Great Lakes, Michigan–Huron (117,400 km^2) and Superior (82,100 km^2), the Great Bear Lake (31,000 km^2), Great Slave Lake (27,000 km^2), Erie (25,700 km^2), Winnipeg (24,514 km^2) and Ontario (18,960 km^2);

In Africa: Victoria (68,870 km^2), Tanganyika (32,600 km^2) and Malawi (29,500 km^2);

In Russia: Baikal (31,500 km^2) and Ladoga (17,891 km^2).

These lakes have been important for fishing and transportation and for the development of humans' boat technology and seafaring experience.

5.2.1.2 Rivers and seas

The Earth's rivers provided the origin of human trading, communication, development and the expansion of human civilisation, see Figure 5.1 and provided access to the sea.

Starting in Africa, its great rivers are the Nile (6,650 km), Congo (4,700 km), Niger (4,180 km), Zambesi (2,574 km), Senegal (1,086 km), Orange (1,432 km) and many others.

In Europe, the main rivers are the Volga (3,530 km), Danube (2,850 km), Don (1,870 km), Dnister (1,362 km), Rhine (1,233 km), Loire (1,006 km) and Rhone (814 km).

In India, the main rivers are the Indus (3,180 km), Brahmaputra (2,900 km), Godavari (1,465 km), Narmada (1,312 km), Mahanandi (858 km), Kaveri (804 km) and Tapi (724 km).

Figure 5.1 Earth's major rivers and seas

In Russia, the main rivers are the Lena (4,294 km), Irtysh (4,248 km), Ob (3,700 km), Yenisei (3,487 km) and Oka (1,500 km).

In Asia, the main rivers are the Yangtze (6,300 km), Yellow (5,464 km), Mekong (4,350 km), Indus (3,180 km), Brahmaputra (2,900 km), Amur (2,824 km) and Liao He River (遼河) (1,345 km).

In North America, the main rivers are the Missouri (3,767 km), Mississippi (3,766 km), Yukon (3,190 km), Rio Grande (3,034 km), Colorado (2,334 km), Arkansas (2,364 km), Red (2,189 km), Columbia (2,000 km), Mackenzie (1,738 km) and Ohio (1,579 km).

In South America, the main rivers are the Amazon (6,400 km), Parana (4,880 km), São Francisco (2,830 km), Purus (3,382 km), Juruá (3,350 km), Paraguay (2,695 km), Orinoco (2,250 km), Rio Negro (2,250 km) and Rio de la Plata (290 km).

In Australasia, the main rivers are the Murray (2,508 km), Darling (1,472 km), Cooper Creek (1,300 km), Murrumbidgee (1,485 km), Lachlan (1,440 km) and Warrego (1,380 km).

All these rivers have all been used by humans for water supply, hunting for fish and riverine animals, for transporting people and for access to the sea.

They continue in use today for trading, some of the most remarkable evidence of the vigour of maritime trade can be seen on maritime app images of ship movements on the rivers and coasts of Europe and of China.

5.2.2 Wind

In Central Asia, in regions where the geography and climate generate strong, predictable diurnal winds, these winds have been exploited from the fifth century for the milling of cereal seeds and for pumping water. The consequential technology developed was vertical axis wind turbines to exploit those winds.

However, in continental East Asia, where winds were more variable and less predictable, in China particularly, from the tenth century, a wide range of lightweight, horizontal axis windmills, using cloth or rattan triangular sails, served similar purposes to those in Central Asia, probably influenced by that Central Asian knowledge.

In Europe, windmills were also exploited for the milling of agricultural products, as in Central and East Asia, but that development occurred later in the eleventh and twelfth centuries, it is unclear whether there was any influence between Europe and Asia.

At the beginning of the nineteenth century, traditional windmills became widespread in the World for the milling corn and other cereals. However, the Industrial Revolution produced several advanced designs, such as that shown in Figure 5.2, developed in the Netherlands, a windmill using a rotating cap, with four adjustable, shuttered blades and complex wooden gearing inside the mill tower to drive the milling machinery.

This European industrialisation of farming technologies was a significant part of the Western Industrial Revolution.

Figure 5.2 A sophisticated Dutch cap-mill from the seventeenth to eighteenth centuries driven by wind to grind corn. Innovations included four blades, cap rotation, gears and stationary milling machinery.

5.2.3 Fossil fuels

5.2.3.1 Coal

Coal, in a wide range of qualities, has been being mined in industrial quantities in Europe and in China since the thirteenth to fifteenth centuries and in the Americas from the seventeenth century.

Production in the early days was largely for burning in the open hearths in the homes and buildings of the wealthy.

However, by the seventeenth century significant coal consumption was being used domestically more widely through society and more commonly in the early metallurgical industries, there as a fuel to smelt bronze, brass, iron and eventually manufacture steel.

From the seventeenth century, a major use of coal became the raising of steam to power machinery at coal mine pit-heads, on railways and throughout an expanding industrial economy world-wide, wherever coal was found.

By the eighteenth century, coal was also being used in Europe, particularly in Germany, for a nascent chemical industry, for heating and separating materials but also as a feedstock for chemicals and medicines themselves.

With the rise in importance of electricity as an Energy Vector at the end of the nineteenth century, a significant role for coal became as steam-raising fuel for coal-fired power stations.

5.2.3.2 Oil

In present-day Iraq, there had been surface leakages of oil since Babylonian times. This oil or tar had been used in Middle East culture as a mortar or cement between baked clay bricks, Fisher [35].

Various civilisations then continued using this oil seepage for construction purposes but then for lighting and low-level heating.

Full-scale exploitation of oil did not really get underway until the mid-nineteenth century when the Scots chemist, James Young, working in Derbyshire observed natural petroleum seepage from coal measures in a mine. From this seepage, he distilled:

- A light thin oil suitable for burning in lamps;
- A thicker oil suitable for lubrication;
- With a realisation that the resource could be valuable for lighting, heating and as an energetic fuel.

Large-scale oil discoveries were then made in the second half of the nineteenth century in Poland, Trinidad, North America, Russia and Iran.

The modern oil industry started from the large-scale exploitation of those resources in the nineteenth century, growing in the twentieth century as other sources in Kuwait, Venezuela, Saudi Arabia and Myanmar were developed, swelling the impact of oil on the Earth's energy scene.

The oil industry developed efficient pipeline systems to supply their products across huge distances and developed countries formed pipeline networks across their territories, between countries and across rivers and seas to form a connected system of supply.

5.2.3.3 Gas and coal-gas

Gas had been a natural by-product of the exploitation of coal measures since the fifteenth century.

However, by the nineteenth century chemical, thermodynamic and energy knowledge was encouraging the exploitation of gas, obtained from either coal or oil measures, for lighting and heating purposes.

Gas also then became a feedstock in Europe for a growing chemical industry, based around coal and gas exploitation.

The use of gas as a fuel, initially from street and house lighting, started to emerge at the end of the nineteenth century and beginning of the twentieth century.

The Industrial Revolution of the nineteenth century led to an increasing need for lighting in mines, factories and eventually in homes. This started in Northern Europe and North America but was by no means limited to those locations.

The supply of coal-gas became a significant Energy Vector in those continents in large cities for lighting and an important demand for coal to be shipped to plants producing the coal-gas in large, heated retorts. These retorts supplied the gas into extensive new pipeline systems, and a valuable side-product of the plants was the coke despatched to iron-works to power blast furnaces as both a fuel and a source of carbon for the steel they were manufacturing.

5.2.4 Other gases

5.2.4.1 Air

The value of air as an Energy Vector is a by-product of the industrial revolution in which new industries require large volumes of air, and other gases, for their industrial processes. Several large-scale companies specialise in the difficult technology of processing, compressing and transporting these gases, including Air Liquide, Air Products, British Oxygen (BOC) and Linde. An example of their development is Air Liquide, founded in France in 1902 for the transport of liquid compressed air and other gases, expanded into Japan 1907, Italy and Spain 1909, Canada 1911 and United States 1916. The reasons for this development are simple, and air and other gases are essential for many industrial processes and by their very nature are difficult to transport. So, either local production was required or reliable transportation to customers in temperature-controlled tanker vehicles or in pressurised containers. Ultimately supply of compressed air by pipeline to huge industrial users began to develop.

5.2.4.2 Steam

Steam is vapourised water, H2O, and the value of steam as an Energy Vector was realised in the late sixteenth to eighteenth centuries, following European scientific developments, particularly from thermodynamics in France in the eighteenth century.

Steam-raising boilers, burning fossil fuels, could heat water to generate steam, which was sometimes stored but then released to do work. This was done in a variety of engineering endeavours, initially associated with coal mining, pumping out-water, lowering down miners and raising their mined material. As the success of these techniques increased, the ubiquity of steam-raising plant to power industrial processes grew and their processes expanded very rapidly, being used all over the World.

Early steam machinery was generally very crude, of poor efficiency and wasteful of fuel. But economic exigencies during the Industrial Revolution and an increasing scientific knowledge, raised the effectiveness of their applications and their efficiencies, as will be made clear later in this chapter.

5.2.4.3 Carbon dioxide

Carbon dioxide, CO_2, is a chemical compound made up of molecules with one carbon atom covalently double-bonded to two oxygen atoms. It is found in the gaseous state at room temperature. In the air, CO_2 is transparent to visible light but absorbs infrared radiation, acting as a greenhouse gas.

The commercial uses of CO_2 rose in the nineteenth century for gasification of drinks and industrial processes such as welding and for the pressing of conformable metal components, its usage also increased as compression and pipeline services emerged.

Pipeline systems within and between countries evolved rapidly as lessons were learnt from the oil and gas industries, which expanded into new Energy Vector industries.

Burning fossil fuels is a primary cause of increased atmospheric CO_2 concentrations and climate change. CO_2 is soluble in water and is found in groundwater, lakes, icecaps and seawater. When CO_2 dissolves in water, it forms carbonates, mainly bicarbonate, HCO_3, which causes ocean acidification as atmospheric CO_2 levels increase.

As the source of available carbon in the carbon cycle, atmospheric CO_2 is the primary carbon source for life on Earth.

Its concentration in Earth's pre-industrial atmosphere since late in the Precambrian has been regulated by organisms and geological phenomena.

Plants, algae and cyanobacteria use energy from sunlight to synthesise carbohydrates from CO_2 and water in a process called photosynthesis, which produces oxygen as a waste product.

In turn, oxygen is consumed, while CO_2 is released as waste by all aerobic organisms when they metabolise organic compounds to produce energy by respiration. CO_2 is released from organic materials when they decay or combust, such as in forest fires. Since plants require CO_2 for photosynthesis, and humans and animals depend on plants for food, CO_2 is necessary for the survival of life on earth.

CO_2 is 53% more dense than dry air but is long-lived and thoroughly mixes into the atmosphere. About half of excess CO_2 emissions to the atmosphere are absorbed by land and ocean carbon sinks. These sinks can become saturated and are volatile, as decay and wildfires result in the CO_2 being released back into the atmosphere.

CO_2 is eventually sequestered, that is long term stored, in rocks and organic deposits like coal, petroleum and natural gas. Sequestered CO_2 is released into the atmosphere through burning fossil fuels or naturally by volcanoes, hot springs, geysers and when carbonate rocks dissolve in water or react with acids.

CO_2 is a versatile industrial material, used, for example as an inert gas in welding and fire extinguishers, as a pressurising gas in air guns and for oil recovery and as a supercritical fluid solvent in the decaffeination of coffee and supercritical drying.

It is also a by-product of fermentation of sugars in bread, beer and wine making and is added to carbonated beverages like beers for effervescence. It has a sharp and acidic odour and generates the taste of soda water in the mouth, but at normally encountered concentrations, it is odourless.

5.2.4.4 Hydrogen

Hydrogen gas was first artificially produced in the early sixteenth century from the reaction of acids on metals. Henry Cavendish, in 1766–1781, was the first to recognise that hydrogen gas was a discrete substance that produces water when burnt, the property for which it was later named, hydrogen, in Greek meaning 'water-former'.

Industrial production of hydrogen has been mainly from steam reforming of natural gas, oil reforming, or coal gasification.

A small proportion of hydrogen gas has been produced using more energy-intensive methods such as the electrolysis of water.

Most hydrogen is used near to its production site, the two largest uses being fossil fuel processing and ammonia production, and this is mostly for the fertiliser market.

Hydrogen can be burned to produce heat or combined with oxygen in fuel cells to generate electricity directly, with water being the only emissions at the point of usage.

5.3 Roads, Great Britain and Ireland

Sophisticated roads were first built in Great Britain by the Romans, see Figure 5.3, who knew it as Britannia. They were designed as a network for military use and created over nearly four centuries, 43–410 AD, when Britannia was a province of the Roman Empire.

Roman army engineers surveyed and built these roads to connect key locations, both strategic and administrative, by the most direct routes. Main roads were constructed of gravel or paved ways, they had bridges constructed in stone or wood, with manned waypoints where travellers or military units could stop and rest.

The next large-scale period of road building in Great Britain and Ireland did not take place until the eighteenth century and the Industrial Revolution.

Figure 5.3 Roman road network in Great Britain, 410 AD

There are four figures who played an important part in the transformation that was about to take place in our islands, John Metcalf, John Loudon Macadam, Thomas Telford and John MacNeill see Figure 5.4.

John Metcalf was a blind Yorkshire man who had worked as a carrier, including a period working with the army, moving guns over difficult ground. In 1765, he won a contract to build a three-mile section of road and applied his military experience, transporting heavy loads, to such good effect, that from 1765 to 1792, he built about 290 km of turnpike road. This was mainly in the North of England, where he became known as the *Father of the modern road*. He believed that a good road needed good foundations and a smooth convex surface that would

(a) (b) (c) (d)

Figure 5.4 Great Britain and Ireland's modernroad, bridge and rail builders. (a) John Metcalf, 1717–1810; (b) John Loudon McAdam, 1756–1836; (c) Thomas Telford, 1757–1834; and (d) John MacNeill, 1793–1880.

Figure 5.5 Following Telford, Metcalf, MacNeill and McAdam's efforts, modern roadbuilding in nineteenth century Great Britain and Ireland

drain easily, see Figure 5.5. He also knew that rain caused the most damage to roads and focused on good drainage, with ditches both sides.

John Loudon Macadam, born in Scotland, showed an interest in road-building at school, but on his father's death, at 14, he moved to the United States. He returned to United Kingdom 13 years later, with the remains of a fortune, made during the US War of Independence. With this, he bought a UK estate and started investigating how to improve roadbuilding. He noticed that gravel-made roads returned to ruts, ridges and potholes when heavy vehicles drove over their freshly spread surfaces. In 1798, he was appointed an agent for revictualling the Royal Navy in the West of England, settling in Falmouth, but in travelling for his work, he noticed that roads were as follows:

> *At once loose, rough and perishable, tedious and dangerous to travel on, and very costly to repair.*

He began experimenting with the stone sizes and the use or road tar to see what improvement could be achieved, becoming the developer of tarmac. His new road-making method had well-drained roads raised above the surrounding land, built from stone broken into cube-shaped fragments. At his own expense, despite local opposition, he built several roads, which were so successful that by 1815 he had been appointed surveyor-general of roads in Bristol, transforming them from dangerous, muddy, rutted, quagmires to even, well-drained surfaces on which carriages could swiftly travel.

Thomas Telford, a Scots civil engineer, established himself as a road and canal project engineer in Shropshire. He designed various infrastructure projects in his native

Scotland, as well as harbours and tunnels. Such was his reputation as a prolific designer of highways and related bridges that he became dubbed the *Colossus of Roads*. As a reflection of his command of all types of civil engineering in the early nineteenth century, he was elected as the first president of the Institution of Civil Engineers. A post he held for 14 years, the town of Telford in Shropshire was named after him.

John MacNeill, a capable Irish civil engineer, was closely associated with Thomas Telford. He gave sound advice for the improvement of Irish main roads during the early part nineteenth century, Figure 5.5. Probably McNeill's most notable project was the development of the Irish railway network.

5.4 Mechanical power technologies

5.4.1 Early steam power

The first applications of modern energy technology were the harnessing of energy from wood and coal for the raising of steam to drive steadily increasing powerful processes, in pumping and locomotion.

There were many pioneers in these fields including the following:

- Pumping:
  Thomas Savery, Thomas Newcomen, Richard Trevithick, James Watt and Matthew Boulton.
- Railways and locomotion:
  Nicolas-Joseph Cugnot, George and Robert Stephenson, Isambard Kingdom Brunel.
- Turbines:
  Charles Parsons [15], George Westinghouse, Gustaf de Laval, Auguste Rateau and Charles Curtis.
- Electricity:
  Michael Faraday, James Clerk Maxwell, Oliver Heaviside, Thomas Edison, Sebastian de Ferranti, Nikola Tesla and George Westinghouse.

The following sections demonstrate how these scientists, engineers and entrepreneurs developed the scientific thought and methods that lead us, increasingly towards the risks of Global Warming, but also towards different ways of energy usage and huge changes in our energy profile.

5.4.2 Reciprocating steam power develops

5.4.2.1 The start

The earliest steam engine in Europe, patented in 1698, was developed by Thomas Savery, a military engineer born in Modbury, Devon, United Kingdom.

He called his machine the Miner's Friend.

Savery heated water in a vessel to make steam and then used a cold-water jet to condense that steam, creating a vacuum, which he then used to draw water up a pipe from a mine into the vessel, Figure 5.6(a). However, the Miner's Friend was

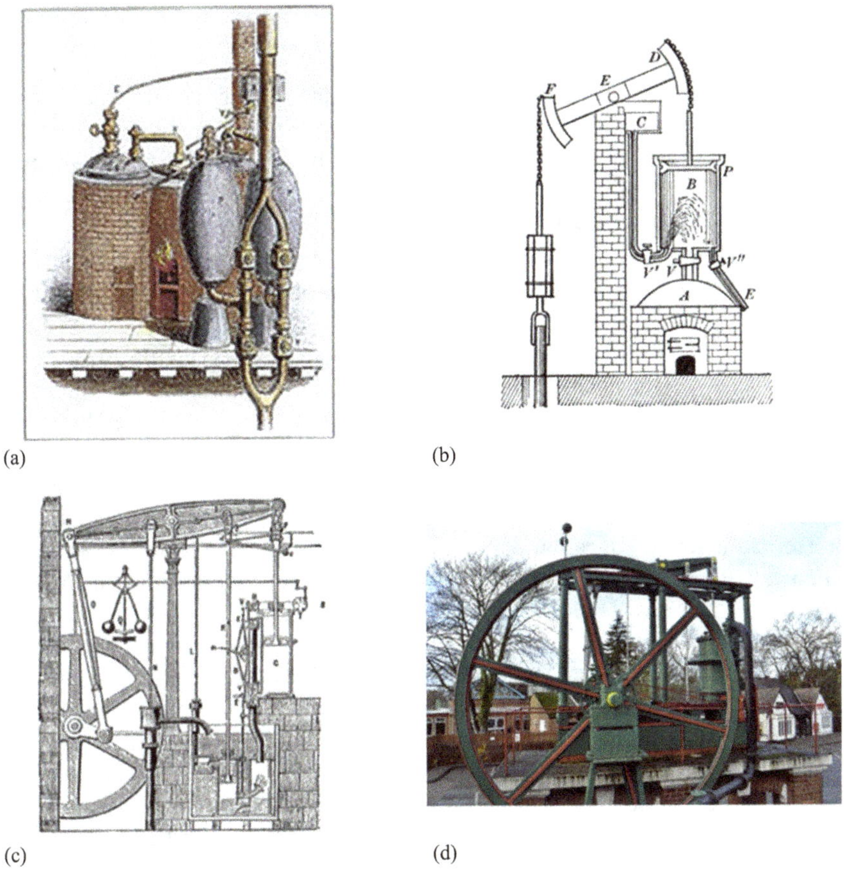

Figure 5.6 Early Steam Engines, from the simple to the complex. (a) Savery's Miners Friend, no moving parts, 1698; (b) Newcomen's mine pump, 1712; (c) Boulton and Watt's beam engine, 1784; (d) a preserved Watt beam engine, Loughborough University, present day.

weak as the height of the water lift was limited by the ambient atmospheric pressure.

Thomas Newcomen, an ironmonger, inventor and lay preacher born in Dartmouth, Devon, United Kingdom in about 1712, developed a reciprocating pump, based upon aspects of Savery's idea. Here steam was raised in a separate boiler, then injected into a cylinder and subsequently condensed to create a vacuum, drawing a piston down to raise water from a mine using a reciprocating pump arm, Figure 5.6 (b). Thus, he contributed the important steam innovation of the condenser.

Neither Savery nor Newcomen had any notion about the efficiency their engines but Newcomen was a well-educated and competent engineer. Richard Trevithick was the son of a Mine Captain between Cambourne and Redruth in

Cornwall, the home of copper and tin mines. Trevithick's innovations were as follows:

- A high-pressure (HP) steam pumping engine for extracting water from the mines;
- And one of the first steam locomotives for hauling product from the mines.

5.4.2.2 Perfecting steam production and flow, James Watt

The lessons from Savery and Newcomen were that steam engines were possible but to be practicable they required a better understanding of thermodynamics and the behaviour of the steam itself.

James Watt, 1784, of Glasgow, United Kingdom, was a gifted instrument maker, inventor, mechanical engineer and chemist, initially working for Edinburgh University, he designed and built a portable gas chamber to facilitate Humphrey Davy's experiments on the inhalation of nitrous oxide.

Watt grew up in the early nineteenth century in the intellectual powerhouse of the Scottish enlightenment, taking place in Edinburgh, Glasgow and Belfast.

He had a sound education and experience of building intricate experimental equipment at one of the foremost technically oriented Universities in Western Europe; here he was the man to sort out the steam engine.

His contributions put steam engine design on a strong scientific footing, requiring a knowledge of basic thermodynamics, a thorough grounding in mechanics, the ability to manufacture effective model prototypes and able to supervise others building full-scale practical machines.

He developed a reciprocating steam engine, Figure 5.6(c) and (d), based upon more scientific principles than either Savery or Newcomen's earlier works, Figure 5.6(a) and (b), and his machines had the following innovations:

- Raising steam in a separate boiler;
- Using that steam to drive a piston in a vertical cylinder;
- Converting the piston's vertical reciprocating motion into rotary motion;
- Rotary motion then drove ropes, shafts or pistons to raise water or material vertically from a mine or source;
- Exhausting steam efficiently from the cylinder and condensing it in a separate condenser. This lowered the outlet pressure, extracting the maximum energy from the steam during the piston's stroke, thereby increasing the engine's fuel efficiency and reducing coal demand;
- The engine's action responded to sophisticated controls, for the admission and exhaust of steam, with a governor to regulate the engine's rotary speed, avoiding the need for constant human supervision.

For the manufacture of his engines, Watt collaborated with a colleague, Matthew Boulton, who owned a factory called the Soho Works in Smethwick, Birmingham. Boulton had the business acumen and drive to develop and sell Watt's inventions to mine owners throughout the United Kingdom.

That partnership, formed in 1775, was to exploit Watt's steam engine patent and began to manufacture the components for steam engines at the Soho Works. The firm lasted over 120 years, albeit being renamed in 1849 'James Watt & Co', but still making steam engines in 1895 when it was sold to a commercial scales manufacturer.

The Watt and Boulton partnership had a unique business approach. They produced the specifications and drawings for their engines, supervised the work to erect them on the customers' premises, but produced almost none of the parts themselves.

Income from the business arose not from the sale price of the pumping engine but from the savings in coal usage, which mine owners using the Boulton and Watt engines achieved. This depended upon the engine thermodynamic efficiency, a key selling point for Boulton and Watt's product but it also depended upon the customer operating the engine in the correct way.

Interestingly, Watt wrote about his visits in the late eighteenth century to Cornish tin mines where his engines had been installed to pump out the workings.

He was trying to assess why the payments the operators ought to be making to Boulton and Watt, from substantial coal savings from using their engines, were not being made.

Rough-hewn Cornish mine-captains, used to operating the earlier, noisy Newcomen engines, were not used to new Boulton and Watt steam engines, which were very much quieter, and operators became used to adjusting Watt's engine valve gear so that they made what they thought of as comforting thumps on each stroke, like a Newcomen engine.

But Watt, after taking a long coach journey from Birmingham to Cornwall to adjust the engines, explained that his engines had to operate smoothly to achieve their thermodynamic gains, which their design guaranteed, and which delivered the coal savings that both parties expected and from which both Bolton and Watt and the mine owners needed to pay for their investment.

It would have been interesting to be able to hear the interchange between Watt and those sceptical Cornish mine-captains, with Watt's precise explanation, in a Scots lilt, of the fine adjustments needed to get the engines into their correct performance regime.

The author remembers another vignette, in the 1950s, as a child seeing a Watt pumping engine still in operation at a North London water treatment works, similar to but much larger than that shown in Figure 5.3(d).

The engine was slow, majestic but almost completely silent, apart from the steady rumble of its bearings and an occasional hiss from its steam control valves.

It is unclear now how Sadi Carnot's thermodynamic ideas influenced Watt in his engines' design. However, it has been proposed that Watt's invention of a separate condenser clarified for Carnot the idea that the flow of heat through an engine was an essential feature of its thermodynamic operation.

The condenser 'made it apparent that a hot and a cold body were necessary if the engine was to work' and this must have been the argument that Watt would have used with Cornish mine-captains.

5.4.2.3 Condensers

Early steam engine condensers were just simple iron containers into which cold water was sprayed to condense the exhausting steam.

Watt designed a more effective condenser, with some similarities to modern designs, converting steam back to water and returning it to the steam-raising boiler as feed-water.

There were many subsequent condenser improvements, including multiple through-tubing, sophisticated tube plates, vacuum pumps to extract the last air from the condenser and special control devices.

Figure 5.11(b) gives us an idea of how sophisticated the condenser could become by 1912. But we must remember that it was Watt who gave us the basic engineered condenser, fulfilling Sadi Carnot's concept of a cold body to extract the maximum energy from a heat engine.

5.4.2.4 Single, triple expansion and compound engines

It was becoming clear from the experiences of Savery, Newcomen and Watt that the potential power of the Boulton and Watt engines could be increased if the full thermal capacity of the steam could be exploited in more than one cylinder.

A compound steam engine is a type where steam is expanded in two or more stages.

A typical compound engine arrangement is where the steam is first expanded in an HP cylinder, then having given up some of its heat and losing some pressure the steam exhausts into one or more larger volume intermediate-pressure (IP) cylinders, finally that steam exhausts into an even larger low-pressure (LP) cylinder before exhausting into the condenser and returning as feed-water to the boiler, see a later development in 1910 for pumping water in Figure 5.4(a), see also Figure 5.9(b).

Multiple-expansion engines employed additional cylinders at progressively lower pressures, to extract further energy from the steam. These techniques, invented in 1781, were first employed on a Cornish beam engine in 1804.

By 1850, compound engines were also being introduced to drive looms in Lancashire textile mills, see a later development in 1935 in Figure 5.4(b).

5.4.3 Tunnels bridges and railways

This introduces us to another eminent British engineer, Isambard Kingdom Brunel, the son of the French engineer Marc Isambard Brunel, married to an English wife, who worked on several important UK projects. These included the development of new methods for manufacturing the tackle blocks for Royal Naval sailing warships. In 1799, Brunel had been introduced to Henry Maudslay, an engineer working for Joseph Bramah, who had recently started his own business, making working models of machines for manufacturing pulley blocks for sailing warships. As a result, Brunel's block-making machinery was installed at Portsmouth Block Mills for the Royal Navy.

Marc Brunel also designed and built the first Thames Tunnel under London's River using a novel shield method and manual labour to dig the Tunnel, where his son Isambard acted as his Clerk of Works. Images of the Brunel family are shown in Figure 5.7.

Figure 5.7 Brunel dynasty: father, son and daughter-in-law. (a) Marc Isambard, (b) Isambard Kingdom and (c) Mary Elizabeth nee' Horsley.

Isambard came to the fore because of experience learnt from his father, his high-quality education, engineering knowledge and drive. At the age of 14, he had been enrolled first at the University of Caen, then at Lycée Henri-IV in Paris, subsequently studying under the master clockmaker and horologist, Abraham-Louis Breguet, who praised Isambard's potential in letters to his father.

Isambard Brunel is now considered one of the most ingenious and prolific figures in British engineering history, see Figure 5.8 for three of his achievements.

In the early part of his life, the use of railways had begun as a major means of transport for goods, influencing his involvement in the railway, including bridge engineering.

In 1833, before the Thames Tunnel was completed, Isambard was appointed chief engineer of the Great Western Railway, one of the wonders of Victorian

(a)

(b)

(c)

Figure 5.8 Isambard Kingdom Brunel's railway achievements. (a) Thames Tunnel, originally designed for horse drawn vehicles, 1843 now carries the Underground; (b) Paddington Station, London, Great Western Railway Terminus, 1854; (c) Royal Albert bridge carrying the Great Western Railway over the Tamar River, Plymouth, 1859.

Britain, running from London to Bristol and later to Exeter, Plymouth and West Wales. It was his vision that passengers would be able to purchase one ticket at London Paddington and travel from London to New York, changing from the Great Western Railway to a Great Western steamship at the terminus in Neyland, Wales.

He surveyed the entire length of the route between London and Bristol himself, with the help of many. Brunel made two controversial decisions:

- To use a Broad Gauge of 7 ft 1/4 in, 2,140 mm, for the track, which he believed would offer superior running at high speeds. His decision to use Broad Gauge for the line was controversial in that almost all British railways to date had used Standard Gauge of 4 ft 81/2 in, 1,435 mm. Brunel said that Standard Gauge was nothing more than a carry-over from George Stephenson's mine railways. He demonstrated, through calculations and trials, that his Broad Gauge was the optimum size for providing higher speeds, greater carrying capacity and stable comfortable rides for passengers. It is interesting to note that he was not alone in considering wider gauges, India selected 1,829 mm and Russia 1,520 mm;
- To take a Great Western route passing north of the Marlborough Downs and then to follow the Thames Valley into London.

The present-day Paddington Station in London was also designed by Brunel and opened in 1854.

An initial group of locomotives were ordered to his own specifications for the Great Western Railway. They proved unsatisfactory, apart from the 'North Star', so 20-year-old Daniel Gooch was appointed as Superintendent of Locomotive Engines to improve designs.

Brunel and Gooch chose to locate their locomotive works in the village of Swindon, a point where the gradual ascent from London turned into the steeper descent to the Avon valley at Bath.

Brunel's achievements ignited the imagination of technically minded Britons of the age, and he became a national icon because of that interest.

After his death, the decision had to be taken to convert his innovative broad to standard gauge to integrate the Great Western Railway into that of other railways throughout the country.

It is interesting to compare the skills of George Stephenson with those of Isambard Brunel. Stephenson was a very expert craftsman and master of intricate steam locomotion detail with a gifted son Robert, whereas Brunel was more of a visionary, with an engineering mind set on the future and the design of long-distance travel by rail and sea.

5.4.4 Locomotion

Key locomotive and railway developments therefore in the eighteenth to nineteenth centuries were, see Figure 5.9:

- Cugnot's 'Fardier à Vapeur' road locomotive (1770);
- Stephensons' 'Rocket' railway locomotive (1829);

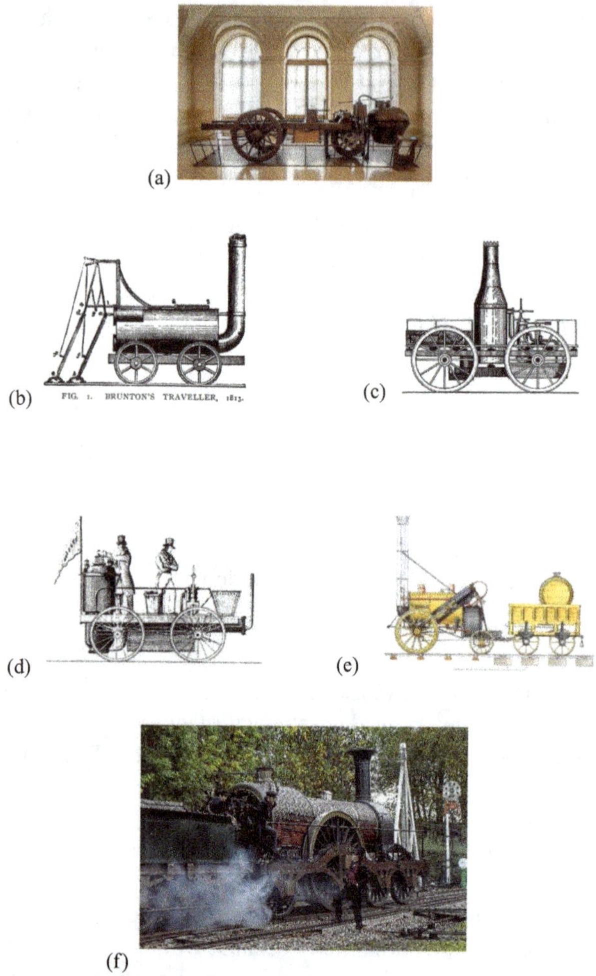

Figure 5.9 Historic steam locomotion development, the 'Rocket' and 'Iron Duke' prove successful, pioneering subsequent world-wide steam locomotive designs. (a) 1770, Cugnot's 'Fardier à Vapeur' 3-wheel road locomotive with vertical boiler and cylinders powering a driving wheel; (b) 1813, Brunton's 'Traveller', with horizontal boiler and cylinders driving walking propulsion mechanism; (c) 1829, Burstall's 'Perseverance', with vertical boiler and cylinder using rolling element wheel bearings. Runner-up at 1829 Rainhill Trials; (d) 1829, Ericsson and Braithwaites' 'Novelty', the world's first tank locomotive, with vertical boiler and cylinder; (e) 1829, Stephensons' 'Rocket', with horizontal boiler and angled cylinders. Winner of 1829 Rainhill Trials; (f) 1842, Daniel Gooch's 4-2-2 'Iron Duke', Great Western Railway with horizontal boiler and cylinders.

- Brunton's 'Traveller' railway locomotive (1829);
- Burstall's 'Perseverance' railway locomotive (1829);
- Ericsson and Braithwaites' 'Novelty' railway locomotive (1829);
- Daniel Gooch's 'Iron Duke' railway locomotive (1842).

5.4.4.1 Initial steam turbine development

The limitations of reciprocating engines gradually became apparent, their slow rotational output speed, large size and consequential costs.

A number of inventors around the World were turning towards the ideas of a turbine, originally conceived by Heron in Alexandria.

Several engineers and entrepreneurs developed patented steam turbine designs for both electrical generation and later ship propulsion markets, including:

- Charles Parsons, United Kingdom, developed in Newcastle-upon-Tyne and patented his design of the reaction turbine, initially for electrical generation but later applied to ship propulsion;
- George Westinghouse, USA, developed reaction turbines under license to Parsons;
- Gustaf de Laval, Sweden, patented his own design of an impulse turbine for high-speed applications;
- Auguste Rateau, France, patented his own design of a pressure-compounded impulse turbine for ship propulsion;
- Charles Curtis, USA, patent attorney to General Electric, formerly Edison, founded the Curtis Marine Turbine Company. This was in conjunction with the Glasgow shipbuilders John Brown & Company, developing in the early 1900s the Brown-Curtis impulse turbine, patenting the design internationally and widely used for electrical generation and ship propulsion.

In later years, the Curtis-Brown turbine outstripped other designs with better performance and lighter weight. At the start of the Second World War, the United States Navy deployed Brown-Curtis turbines in their warships, with better performance than Royal Navy ships, equipped with heavier turbines.

Parsons had developed the first modern practical reaction steam turbines, see Garrett Scaife [15]. Because of his turbines' high-speed, their first application in 1887 was to drive 2-pole DC commutator electric generators shown in Figure 5.10(a).

However, Tesla's experience in the United States on AC motors and generators leads, from 1899 onwards Parsons moved towards lower speed turbines driving 4-pole or subsequently 2-pole AC generators, Figure 5.10(b), developing from 1930 as shown in Figure 5.10(c).

An interesting aspect of Charles Parsons' life in Newcastle, Figure 5.11, was that his wife, Katharine, worked closely with him on engineering projects, especially during the steam turbine development period. She had no technical training but a tough approach with good organisational skills, becoming an influential contributor to C A Parsons Ltd and an advocate for women in engineering.

During the First World, she became closely involved in managing the female workforces at both Works, at Heaton and nearby Wallsend-on-Tyne, as the

(a)

(b)

(c)

Figure 5.10 *Development of Charles Parsons steam turbines for driving electrical generators. (a) Early 10,000 rev/min DC reaction steam turbine, 1887 driving 2 kW DC generator. (b) A 1,500 rev/min reaction steam turbine, 1899 driving a single phase 1 MW, 4 kV, AC generator, Elberfeld, Germany. (c) Erection of 3,000 rev/min reaction steam turbine generator, Heaton Works, 1930 left HP turbine; centre 2 stage LP turbine on crane; right 2-pole generator rotor.*

(a)

(b)

(c)

Figure 5.11 Parsons family members involved in engineering. (a) Sir Charles, (b) Lady Katharine and (c) Daughter Rachel.

factories replaced the men who had joined the Services with young women to do the work.

Having seen the effective deployment of women in engineering work during the War Katharine was critical of their removal form their factories under the Restoration of Pre-War Practices Act 1919. In a widely publicised speech in that year on '*Women's Work and Shipbuilding during the War*', she deplored the way women had been required to produce the '*implements of war and destruction*' but were denied '*the privilege of fashioning the munitions of peace*'.

Charles and Katharines' daughter, Rachel, passed the Mechanical Sciences Tripos Part 1 at Newnham College, Cambridge University before the First World War, at a time when female Cambridge undergraduates could not take a degree. She contributed during the War as a Director at Heaton Works, becoming a member of the National Council of Women, in which she campaigned for female access to technical schools and colleges.

Katharine Parsons' war-work led her to become the first Honorary Fellow of the North-East Coast Institution of Engineers and Ship-builders.

Rachel Parsons also became a member of the Royal Institution of Great Britain. Rachel and Katharine then became founders of the Women's Engineering Society in United Kingdom in 1919.

Rachel was a wealthy, eccentric and reclusive but forgotten feminist pioneer, murdered in her Suffolk home in 1956, shocking the country.

However, the results of Rachel and her mother's remarkable technical contributions began to emerge. Their family has made a very significant contribution to the development in Ireland and Great Britain of technical skills and the opening of engineering and other technical professions to women.

Competitors in Europe and USA, Westinghouse, de Laval, Rateau and Curtis, were developing their own turbines, Figure 5.12, showing the differences between reaction and impulse turbines, the former being faster, the latter slower:

- A reaction turbine is composed of moving nozzles alternating with fixed nozzles, such as the Parsons turbine.
- An impulse turbine is composed of buckets alternating with fixed blades or nozzles, such as the Brown-Curtis turbine.

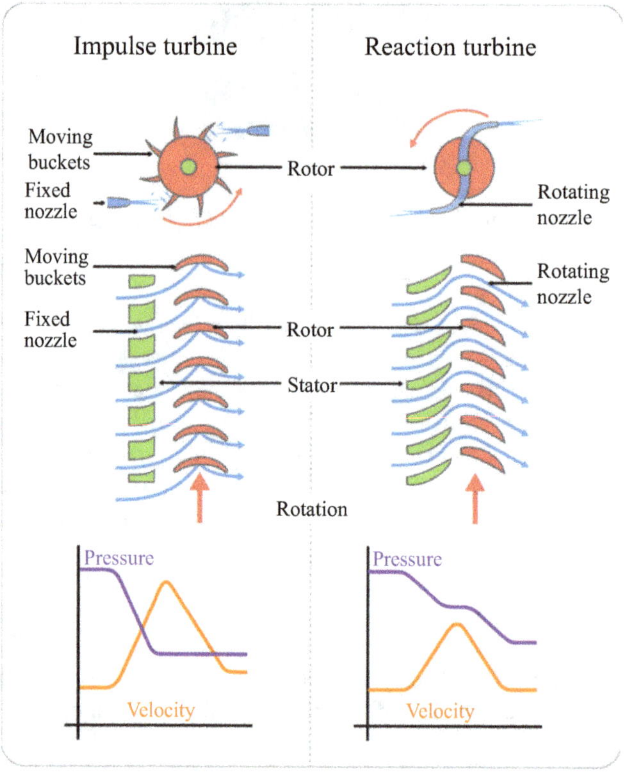

Figure 5.12 Examples of reaction and impulse steam turbine operation [15]

The nozzles of a turbine appear like blades, but their profiles converge near the exit. This convergence results in a steam pressure drop and velocity increase as steam moves towards the moving blades. The turbine blades move due to both the impact of steam on them and the reaction due to the high velocity of the steam exiting them. As experience grew with these different types of turbine complex, combinations began to be employed to exploit their particular advantages.

5.4.4.2 Reaction, impulse and Brown-Curtis turbines

Modern steam turbines adopt several blade types and arrangements to extract the required energy from steam, so the distinction among reaction, impulse and Brown-Curtis types of blading is now less distinct and all maybe employed in the same turbine.

Figure 5.13 shows a section through a modern marine steam turbine, from a Royal Navy frigate, in which the author served in the Persian Gulf, it incorporated on one shaft a combination of:

- The ahead turbine, with a Curtis double wheel followed by 11 successive reaction turbine wheels;
- The astern turbine, a Curtis double wheel followed by a single impulse turbine wheel.

5.4.5 Further reciprocating steam engine development

5.4.5.1 Advanced reciprocating steam power

This section shows the development of the reciprocating steam engine beyond the pioneering work of Watt and early locomotives.

It demonstrates the power of the reciprocating steam engine for driving low speed, high torque machines, enabling the development of the railways, water-borne propulsion, mine water and sewage pumping, all applications that require a large tractive effort.

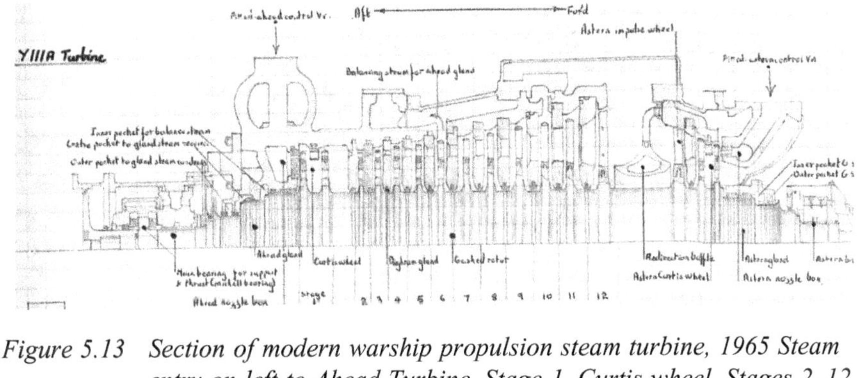

Figure 5.13 Section of modern warship propulsion steam turbine, 1965 Steam entry on left to Ahead Turbine, Stage 1, Curtis wheel, Stages 2–12, 11 successive reaction nozzles and blades, Steam entry on right to Astern Turbine, Brown-Curtis wheel following an impulse wheel

Therefore, the reciprocating steam engine continued to survive, from the eighteenth century into the twentieth century, growing in size and efficiency in specific appropriate industries, Figure 5.14.

But there was always scientific effort to identify whether combinations of reciprocating, turbine and other technologies could achieve higher torques, better efficiency or higher speeds.

(a)

(b)

Figure 5.14 Sophisticated, large power, modern design reciprocating steam engines. (a) Vertical Triple Expansion Steam Engine, 108 HP at 39 rev/min, most common waterworks pumping engine, based on Watt's principles, 1910. (b) Horizontal Cross Compound Mill Steam Engine, 80 HP at 120 rev/min, Manchester College of Technology for teaching purposes, 1935.

5.4.5.2 Improved locomotive applications

It could be argued that the most beautiful steam locomotives, the 'Mallard' Figure 5.15(a), based upon reciprocating technology produced in Britain, where the 'Rocket' had been the pioneer, see Figure 5.6(e).

But there can be no doubt the United States the most powerful locomotives, see Figure 5.15(c). Steam locomotive thermal efficiencies were notoriously low <4% but France produced the most thermodynamically efficient locomotives of the type shown in Figure 5.15(b), which were measured and probably achieved efficiencies better than 6%.

Many locomotive efficiency lessons were learnt in Europe and the USA from French engineers' clear thermodynamic understanding of their locomotives' performance. Some of those lessons, developed elsewhere, were learnt eventually by locomotive designers in United Kingdom in the late 1930s–50s.

Significant elements of steam locomotives' uninspiring efficiency were their use of a reciprocating motion, the lack of a condenser and consequential open-loop water system.

There were several attempts in United Kingdom, Sweden and Germany at resolving this situation, see [16].

The Reid-Ramsay design in United Kingdom of 1909 was a turbine-electric tank locomotive using a steam turbine with condenser to drive a generator, supplying DC electricity to four traction motors, between the trailing bogies at front and rear of the locomotive.

The literature says this turbine was an impulse type, but the author suspects it was more likely to have been a Parsons reaction turbine.

The DC electrical transmission proved unreliable, and the locomotive was subsequently rebuilt to be a turbine-mechanical locomotive, driving through gears and known as the Reid-McLeod locomotive, Figure 5.16(a).

In 1921 Ljungström, an eminent Swedish steam engineer, developer of the impulse turbine, used his design for a new locomotive solution with transmission from the turbine driving the axles via gears, achieving some success in Sweden, Figure 5.16(b).

The locomotive towed its own condenser, clearly visible in the photograph, to return steam from the turbine back to feed-water for the boiler.

United Kingdom, Swedish and German railways followed suit from the Reid-McLeod and Ljungström experiences, developing turbine locomotive designs, both smaller and larger, for passenger or freight traffic.

However, the turbines, gearing and condenser proved too complex for modern steam locomotive maintenance teams and turbo-locomotives were not an operational success.

World railways continued operating traditional reciprocating steam locomotives without condensers, with resultant low efficiencies, right up until the 1960s–70s. Only one or two countries, pioneered by French experience, starting to improve traditional steam locomotive designs.

(a)

(b)

(c)

Figure 5.15 Reciprocating steam locomotives of the twentieth century. (a) United Kingdom, London North Eastern Railway, 'Mallard', 4468, 4-6-2, holder of the World speed record for steam, 126 mph (203 km/h) 1938. (b) France, SNCF, 4-8-2, 241 P17 Steam Locomotive. (c) USA, Union Pacific Railroad, 'Big Boy', 4014, locomotive with 2 articulated power units, 4-8-8-4 wheel arrangement.

(a)

(b)

Figure 5.16 Steam turbine-electric condensing locomotives. (a) Reid-McLeod steam turbine-electric condensing locomotive from North British Locomotive Works, Glasgow, 1924, (b) Ljungström steam turbine condensing locomotive at Stockholm Central Station, 1922.

However, during First World War and Second World War, there had been significant improvements in diesel and electric traction performance.

These improvements manifested themselves, in the 1950s–90s, through higher power/weight ratios and lower maintenance, with resultant growth first in diesel locomotive train traction then in electric multiple-units, proliferating from 1950 to 2020 on railways in Europe and the Far East, particularly around large cities.

From 1970 to 2010 diesel or electric locomotive haulage of heavy freight trains became the norm.

An unstoppable avalanche of diesel and electrical improvement made the steam locomotive redundant.

5.4.5.3 Ship applications

The application of steam power to ship propulsion started in the early part of the nineteenth century, initially relying on simple single expansion reciprocating steam engines to drive large paddle wheels and the early applications were for vessels working as harbour tugs. But these small paddle steamers quickly expanded into applications on larger passenger vessels on European and American rivers and there was an evolution into more thermodynamically sound steam engines with higher efficiencies.

Key developments in modern ship propulsion in the nineteenth and early twentieth century were as follows:

- Isambard Brunels' 'Great Western' (1838), 'Great Britain' (1843) and 'Great Eastern' (1858) all with reciprocating steam engines;
- Harland and Wolffs' Olympic Class vessels, including the TITANIC (1912), which combined high performance triple expansion steam reciprocating engines with the modern steam turbine;
- Steam turbine-driven ships from France, Germany, United Kingdom and USA primarily for transatlantic mail and passenger services.

At the turn of the nineteenth to twentieth centuries, competition between steam reciprocating engines and Parsons' turbine innovations challenged the shipbuilding industry.

The Olympic Class of passenger of liners, conceived in the early twentieth century, were built by Harland and Wolff in Belfast, and consisted of three ships, each of about 46,000 tonnes, including the ill-fated TITANIC, Figure 5.17(a).

These ships adopted some new turbine engineering practice but also incorporated evolved older, triple expansion, steam machinery design features.

The ships were equipped, for the time, with an unusually efficient coal-fired, steam propulsion arrangement, shown in Figure 5.17(b).

This continued to incorporate reciprocating combined turbine machinery technology, designed to reduce the coal tonnage on board the ship, maximising space for the lucrative passenger market.

Each ship could carry about 3,500 people in total, of whom about 2,500 were passengers, the ship's coal-fired propulsion required up to 900 men to handle the fuel, stoke the boilers and operate the ship's machinery.

Steam for these huge ships was raised by burning hard coal in 29 Scotch boilers, feeding their steam output at a pressure of 14.8 Bar, 201 °C to 2 parallel, reversible triple expansion reciprocating engines and then one centre-line LP reaction steam turbine.

The huge reciprocating steam engines drove each of the port and starboard propellers, whilst the steam turbine drove a centre-line propeller, see Section 5.4.2.

The use of triple expansion reciprocating steam engines was a conventional choice for the time, as other large liners were increasingly being powered by steam turbines, following Naval experience to increase ship speed.

The Olympic Class's engines were the acme of reciprocating design for the time. These were powerful, reliable and well-understood engines, whilst the addition of a steam turbine enabled the propulsion system to extract the latent energy of steam exhausted from the two reciprocating engines, providing a steady cruising thrust to the vessel.

(a)

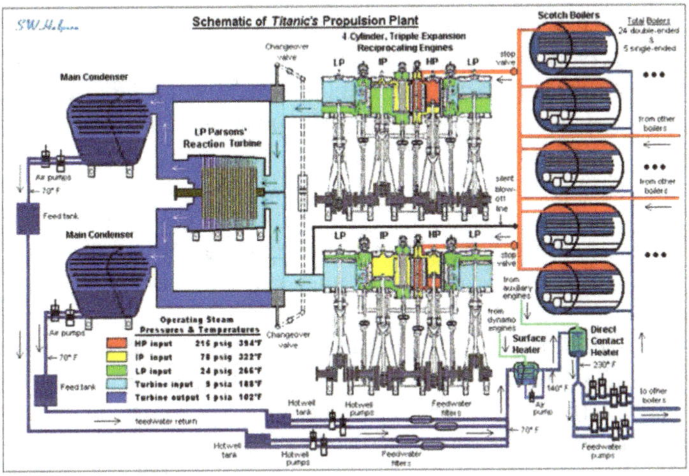

(b)

Figure 5.17 RMS TITANIC, White Star Line. (a) TITANIC at Sea, (b) Boiler and Engine Room Schematic, showing Boiler Room, Port and Starboard engines, 4-cylinder, Triple Expansion Reciprocating and a centre-line engine, single cylinder, Parsons' LP Reaction Turbine. Courtesy: S W Halpern.

Referring to Figure 5.17(b), therefore, Olympic Class ships passed steam from their boilers through:

- Two triple expansion reciprocating engines, driving Port and Starboard propellers, with steam flowing through:
  - One HP cylinder for each engine;
  - Exhausting into one IP cylinder;
  - Followed by two parallel LP cylinders;
- Port and Starboard reciprocating engines then exhausted their steam into a third LP engine, the reaction turbine driving the non-reversible Centre-Line Propeller;

- Finally, the steam from all three engines exhausted into two huge Parsons' condensers, cooled by seawater, where the steam was returned to feed-water and residual heat dumped into the sea. The condenser then returned condensed feed-water to the boilers.

For a ship of the time, this was a complex but very modern steam propulsion system.

The engines represented a well-designed thermodynamic arrangement to extract maximum energy from the coal burnt with high efficiency.

There was a minimum wastage of heat from the propulsion machinery, including the novel addition of a reaction turbine for the centre-line propeller, to extract the last of the steam's energy before it was condensed and returned to the boilers' feed.

This centre-line propulsion provided the ship with highly efficient cruising power.

The overall thermal efficiency of the ship's propulsion machinery was probably of the order of 10%–14%, very much better than the thermal efficiency of any steam locomotive in operation at that time, or since.

The machinery is of interest to us now because it shows that, even in early twentieth century industry, increasing attention was being paid to the importance of efficiency for coal-fired machinery, derived from timely thermodynamic knowledge, strongly influenced by Carnot's ideas, albeit aimed at improving the vessel's overall profitability.

The inclusion of a Parsons' steam turbine was relatively new for passenger ships, at that time, reflecting his experience initially developing steam turbines for application to high speed such as Royal Navy ships, based on his first small prototype TURBINIA.

In fact, before developing, his steam turbine Parsons had successfully developed a four-cylinder, high speed, rotating reciprocating engine driving a dynamo, which won him a 1885 Inventions Exhibition prize, see below.

The loss of life at TITANIC's sinking in 1912 was about 1,500, from an estimated 2,224 people on board. A disproportionate number of those deaths were from the lower decks, particularly amongst boiler and engine room staff, responsible for keeping those immense boilers, reciprocating and turbine engines working.

However, the Olympic Class of ships was a successful demonstration of modern ship construction and steam propulsion, with many excellent thermodynamic lessons learnt from earlier steam-powered ships, including the triple expansion reciprocating engines, the LP steam turbine and the condensers.

Subsequently, some of these developments were absorbed into the designs for many of the world's more recent coal-fired power stations for the supply of electrical energy.

However, later understanding of TITANIC's fate suggests that the ships' hull integrity, rudder design and life-boat provision of the Olympic Class certainly needed improvement.

The author served as a Naval Officer on board frigates in the Persian Gulf and this vignette, in 1965, is that these ships were powered by some of the technologies described in this chapter.

At that time, in Bahrain, the Royal Navy was still using frigates for piracy patrols with steam reciprocating engines, like but smaller than those fitted to RMS TITANIC, Figure 5.17(b). These engines delivered a calm, smooth 18 knots to their ship in pursuit of pirate dhows in the straits of the Persian Gulf.

The author's own ship, a relatively new frigate, HMS ASHANTI, was powered by a combination of a modern steam turbine and an ultra-modern gas turbine, called the Combined Steam and Gas turbine or COSAG arrangement, and was able to operate at higher speeds.

Figure 5.10 shows a section through that ship's steam turbine, which when compared to the turbines in the earlier TITANIC, Figure 5.17, demonstrating from 1911 to 1965 significant changes, in terms of engine design and power-to-weight ratio.

The gas turbine gave our frigate a power boost, which could accelerate the ship very quickly away from the jetty, when needed to put to sea rapidly achieving a speed of up to 25 knots to chase and catch faster dhows, powered by modern Evinrude diesels.

There were some architectural similarities between HMS ASHANTI's propulsion arrangement and that pioneered in RMS TITANIC, Figure 5.17(b), but there were greater similarities to the forthcoming Combined Cycle Gas Turbine (CCGT) power stations introduced in the 1990s, see Figure 6.8.

5.5 Electrical power technologies

5.5.1 Electrical innovators

Magnetism was beginning to be understood in the sixteenth and seventeenth centuries but the electrical revolution started in the eighteenth century. That development depended on scientific innovation, engineering knowledge and business acumen to develop the ideas being presented by scientists and engineers in the nineteenth century.

These ideas coalesced from many different countries, and important contributions came from those described below:

- Michael Faraday, United Kingdom (1791–1867)
  Self-taught natural philosopher, one time assistant to Sir Humphrey Davy, Faraday was a pioneer in the study of electrochemistry and laid down the principles of electromagnetism;
- Ernst Werner von Siemens, Germany (1816–92)
  Businessman and founder of the German electrical engineering company Siemens and Halske surviving today as Siemens AG, a German multinational conglomerate corporation and the largest industrial manufacturing company in Europe, headquartered in Munich with several foreign branch offices.
- Sir William Siemens, Germany and United Kingdom (1823–83)
  Younger brother of Ernst von Siemens, the fourth of eight surviving family sons. His primary interests were electric telegraphy and lighting engineering who founded the electrical company Siemens Brothers in United Kingdom.

At the end of First World War, Siemens Brothers and Company Ltd became English Electric being subsequently in 1955 acquired by Associated Electrical Industries (AEI), merged with GEC in 1968 and ultimately becoming a part of the US electrical giant General Electric (GE), originally founded by Edison.

- William Thomson, United Kingdom (1824–1907)
  British mathematician, mathematical physicist and engineer born in Belfast. Professor of Natural Philosophy at University of Glasgow for 53 years. He did important work on the mathematical analysis of electricity and the formulation of the first and second laws of thermodynamics. He also helped to unify the emerging discipline of physics in its contemporary form. Subsequently Lord Kelvin;
- James Clerk Maxwell, United Kingdom (1831–79)
  Scots mathematician and scientist, Professor at Marischal College, Aberdeen University, then Professor at Cambridge University. Finally he was Professor at King's College, London University, probably the most productive period of his career. Responsible for the classical theory of electromagnetic radiation, the first to describe electricity, magnetism and light as different manifestations of the same phenomenon;
- George Westinghouse, USA (1846–1914)
  American engineer and entrepreneur based in Pennsylvania developed the railway air brake first, receiving his first patent at the age of 19. He founded the Westinghouse Electric Corporation as a manufacturing company in 1886 and pioneered an electrical industry. In the early 1880s, he saw the potential for using alternating current AC for long-distance electric power distribution, putting his resources into developing and marketing AC electrical networks, who founded the Westinghouse Electric Corporation, in direct competition with Thomas Edison, who marketed DC power distribution. In 1997, the Westinghouse Power Generation Business Unit, headquartered in Orlando, Florida, was sold to Siemens AG.
- Thomas Edison, USA (1847–1931)
  American inventor and businessman, who developed devices in the fields of electric power generation, mass communication, sound recording and motion pictures. His inventions included the phonograph, the motion picture camera and early versions of the electric light bulb. These have all had a widespread impact on the modern industrialised world.
  The Edison Electric Company was formed in 1892 by the merger of the Edison General Electric and Thomson-Houston Electric Companies to eventually become part of the US electrical giant General Electric (GE);
- Oliver Heaviside, United Kingdom (1850–1925)
  Self-taught English mathematician and physicist who invented a new technique for solving differential equations, independently developed vector calculus, and rewrote Maxwell's equations in the form which is commonly taught and used today [6]. A prolific writer of electrical scientific material, an unusual yet effective scientific thinker, his vision of the mathematics and usage of electromagnetic principles is at the forefront of present-day teaching;

- Nikola Tesla, Serbia and USA (1856–1943)
  Serbian–American inventor, electrical and mechanical engineer and futurist. Best known for his contributions to the design of the induction motor AC generator and his greatest contribution, the modern alternating current (AC) electricity supply system. He worked with George Westinghouse designing AC generator, transformer and motor products for the Westinghouse Electric Corporation;
- Sebastian de Ferranti, Italy and United Kingdom (1864–1930)
  Son of an Italian father and English mother, educated at University College London, a talented electrical engineer who worked for the Siemens Brothers in London, designing electrical devices including an arc lamp for street lighting and an electrical generator, designed with William Thomson's help, subsequently patented as 'the Ferranti Dynamo'. His UK electrical engineering and equipment firm, Ferranti Ltd, operated for over a century from 1885 until 1993, finally Ferranti International PLC went bankrupt, following the acquisition in 1987 of International Signal and Control, a suspect US defence contractor.

This UK-centric snapshot of the scientists, engineers and entrepreneurs, who developed early practical electrical principles, includes voltage generation, power production, energy transmission, electrical conversion to mechanical output and communications.

These principles were gradually conceived and put into operation, around the World, in the nineteenth and twentieth centuries from 1770 to 1940, with all their technical, social and innovative possibilities and their environmental impacts.

Scientists and engineers in various countries then evolved these electrical principles, putting them to work in their own industrial conditions and environments.

5.5.2 Early power stations

Charles Parsons' first foray into electrical engineering had been in 1881, before his work on turbines, when he had designed a high-speed, reciprocating rotary steam engine to drive a DC electric generator, in collaboration with Charles and Henry Kitson, Figure 5.18.

The generator, built by Brush, was to supply a DC lighting source. This machine won a gold medal at the London Inventions Exhibition, 1885; however, the technology did not make an electrical breakthrough. The initial key developers of the forthcoming world electrical power breakthrough, with their associated early innovations, were as follows:

- Thomas Edison, Pearl Street DC Coal-fired Power Station, steam reciprocating engines, 0.8 MW, New York (1882–86);
- Sebastian de Ferranti, Deptford AC Coal-fired Power Station, steam reciprocating engines, 30 MW, London (1885–90);
- Nikola Tesla and George Westinghouse, AC Hydro-Electric Power Plant, hydraulic turbines, 11.1 MW, Niagara Falls (1895).

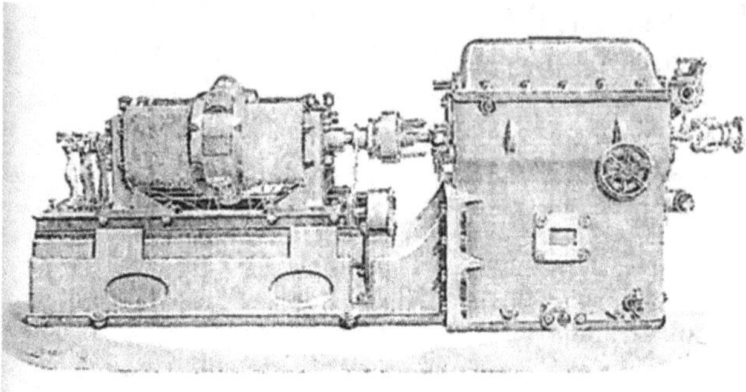

Figure 5.18 On left, the first high-speed steam reciprocating epicycloid steam engine. On right, driving a Brush DC generator, 1881.

Their earliest power stations were either coal-fired or water-powered. The most influential developments were in the USA and United Kingdom between 1880 and 1900 but soon spread to other European countries and around the World.

The World's first electrical power station was Thomas Edison's Pearl Street DC power station in New York, United States, Figure 5.19(a) and (b), in 1882.

This was followed by Sebastian de Ferranti's AC power station in Deptford, London, Figure 5.19(c) and (d), in 1889.

Both power stations employed reciprocating steam engines, Edison favouring several small machines, while de Ferranti used two huge machines that can be seen in Figure 5.19(d).

However, there had been many other electrical developments in many other countries. For example, the installation by Siemens of electric lighting in the Summer Palace, YiHeYuan (頤和園) of the Empress Dowager CiXi, in Beijing (北京), China, in or about 1900.

The engineer Edison's inventive spirit, in the United States, appreciated the applicability of his electrical inventions to American city's commercial environment. Therefore, he concentrated on supplying electric power to the wealthy centre of New York, profiting from his DC generation and lighting inventions.

However, he had not anticipated the difficulties that could arise from transmitting DC power over longer distances than those encompassed by the wealthy financial districts around Pearl and Wall Streets.

The effects of the rapid expansion in New York of technical developments in electrical, telegraphic and telephone technologies can be seen in the aerial chaos shown in Figure 5.12, highlighted by the low temperature conditions at the time.

On the other hand, in the United Kingdom in 1889, the engineer Sebastian de Ferranti took a different route. For reasons of coal supply, he decided to build his London coal-fired power station in Deptford, about 13 km down-stream the River Thames from the City's wealthy centre, where the electricity could be sold at a good price.

Coal barges could come alongside the Deptford wharf to unload their coal directly into his power station furnaces.

(a) (b) (c) (d)

Figure 5.19 Early Central Power Stations in United States and United Kingdom, employing steam reciprocating engines. (a) Edison's coal-fired DC, Pearl Street Power Station, 600 kW DC, first central power plant New York, United States, 1882; (b) horizontal axis DC generators inside Edisons's Pearl Street Power Station, furnaces below generators; (c) Ferranti's coal-fired AC Power Station, 2.2 MW AC, Deptford, first electrical power station London, United Kingdom, 1889; (d) large horizontal axis AC generators inside Ferranti's Deptford Power Station, furnaces to right of generators.

Ferranti therefore needed a transmission line to transmit electrical energy from Deptford up to London. He proposed using the track-bed of the London to Greenwich Railway as the place to lay a high voltage AC cable.

At the power station, he opted to use as the power source horizontal axis, slow-speed, multi-pole AC generators, of his own unique design, driven by large steam reciprocating engines, fired by coal-fired furnaces, Figure 5.19(d).

He then used transformers to elevate the generator output voltage of 1,200 V to a 10 kV HV AC cable voltage, more suitable for the long-distance transmission from Deptford to London.

This higher cable voltage lowered the transmitted current, reducing transmission losses.

Electrical power was then transmitted, by this early cable transmission system along the railway, from Deptford to the Grosvenor Gallery in the centre of Town, where transformers reduced the voltage from 10 kV down to 50 or 100 V AC, safe usable voltages suitable for buildings and customers close to the substation, who primarily used the power for lighting their houses.

Meanwhile in Europe, electrical science inspired several engineers to start developing a wide range of electrical patents, covering generators, transformers, transmission systems and switchgear.

These included Nicola Tesla, from present-day Serbia, who applied his inventive mind to the development of an AC induction motor. He emigrated to the USA to continue his work on AC electrical machinery, including the induction motor and AC generator. He did not use mathematics or symbols to design his products but his incredible imaginative powers to envision their operations in his head.

This author confirms that he found the descriptions of induction machine operation, based on the mathematical approaches of Gabriel Kron, devoid of physical meaning, unsuccessfully forced on him as a university student. He relied more heavily on the physical explanation of Nicola Tesla and the equivalent circuits of Philip Alger. However, he realises now that Kron's methodology enabled us, in recent years, to transfer the modelling dynamics of AC machinery onto the digital computer.

Nicola Tesla's good fortune was to team up with George Westinghouse, an entrepreneur and engineer, capable of exploiting Tesla's unique knowledge and understanding.

Westinghouse had started his career developing air-braking systems for railroads. As they expanded in the United States, he became wealthy and then turned his attention to power generation. He focussed his business acumen on developing an electric power network for the whole USA, where his vision was to develop AC generation with HV AC transmission, to minimise losses, so that his products could power huge, long-distance continental AC electrical networks.

This was in contrast with Edison's more restricted approach in New York, supplying LV DC over short distances to wealthy New York customers and businesses.

Westinghouse's fascination with the long-distance supply of AC power was enabled by the many patents that he bought up from famous engineers in Europe and the United States.

Tesla and Westinghouse collaborated on the development in 1895 of the world's first hydro-electric power station, Figure 5.6, using the Niagara Falls to power vertical axis, slow-speed, multi-pole, AC generators and transformers of

Tesla design to supply New York, over 660 km away, delivering power via an early HV AC overhead transmission system.

The electrical partnership between Tesla and Westinghouse has some similarities with that earlier steam partnership between Watt and Boulton, combining technical knowledge with a hard-headed business understanding.

The application of water to power for humans had started with the water wheel. By the end of the nineteenth century, the hydro-electric power station had superseded the earlier watermill, Figure 5.20(a).

By the late nineteenth century waterpower, it progressed to hydro-electricity. These developments substituted water for direct work to water for generating electricity, then using electrical networks to channel power from the source to customers, where local work could be done.

This is the intrinsic value of an electrical transmission and distribution system for parcelling out work.

Long-distance electrical power transmission continues today around the World, exemplified by Figures 5.19, 5.20 and 8.8, where the following comparisons can be made:

- Figure 5.20, Tesla and Westinghouse's Adam's Niagara Falls Power Plant, built 1895, with a capacity of 37 MW hydro-electric plant supplying Buffalo, 20 miles distant, with AC electricity. The power plant was steadily increased in size until it eventually transmitted power at 20 kV to New York, 660 km away.

  The present Robert Moses Hydro-electric Power Plant, Lewiston, New York, near Niagara Falls, built 1961 with 13 generators has a capacity of 2,525 MW;
- Figure 5.21, a small modern high-head electrical generating plant employing an impulse turbine, Pelton wheel, delivering 1 MW from a high-head hydro-electric source;
- Figure 8.8, China's Three Gorges (San Xia Da Ba) Hydro-electric Plant, built 2011, with a capacity of 22,500 MW feeds an AC substation in Yichang 43 km away, where power is converted to HV DC and despatched in 4 links at 500 kV to the city of Shanghai (上海), 1,150 km away, see Section 8.8.

These are the important lessons to learn from the comparison of hydro-electric and coal-fired electrical plants:

- Hydro-electricity, the water fuel is free but is channelled into a dam and power station, the construction of which entails very high capital costs, but with an operational life as high as 80–120 years, so capital costs are amortised over a long period.

  But there are likely to be seasonal variations in water throughput, therefore, both suppliers and consumers have to adjust to this variability;
- Coal-fired plants, the fuel is costly to mine but cheap to transport to power stations close to consumers; therefore, suppliers can more readily meet varying consumers' demand.

(a)

(b)

(c)

Figure 5.20 Early waterpower and an US' first AC hydro-electric power station. (a) Early eighteenth century Watermill; (b) Adam's Power Plant Niagara Falls, United States, Machinery designed by Tesla, built by Westinghouse, total output 37 MW (1895); and (c) Inside Adam's Niagara Falls Power Plant, United States, showing three Tesla design vertical hydroelectric AC generators, each rated at 3.7 MW (1895)

The mining and burning of coal as a fuel carries a heavy environmental impact and power stations' thermal efficiencies cannot achieve better than 40%, in many practical cases being no more than 20%.

However, the capital cost of coal-fired power stations is not as high as nuclear or hydro plant and their operational life can be as high as 40–50 years.

Figure 5.21 Waterpower, small modern AC hydro-electric power station, single high head impulse turbine, Pelton wheel driving an AC electric generator

5.5.3 Energy networks

5.5.3.1 Electrical transmission

Early electrical transmission systems include examples shown of New York in Figure 5.22 in the 1880s.

First, Figure 5.22(a) shows the chaotic state of early electrical distribution, and the second, Figure 5.22(b) shows the impact of electric street lighting on a prosperous area of Manhattan.

5.5.3.2 London's urban rail network

In the mid-nineteenth century early railway systems in Great Britain and Ireland, Europe, India and the Far East, evolved rapidly into extensive networks.

As the railway revolution spread, more thought then went into network design so that cities were served more comprehensively.

Great Britain was at the forefront of this early railway expansion.

At that time, London was one of the World's largest cities, and for reasons of planning, main-line railways were forbidden to enter the centre of London, within a roughly rectangular box 2 miles by 4 miles around, see Figure 5.23, being kept outside to keep the city unencumbered with railway lines and free of their smoke and noise.

The terminus stations for these new main-line railways were Liverpool Street, Kings Cross, St Pancras, Euston, Marylebone, Paddington, Victoria, Charing Cross and Cannon Street, all lying just on the edge of that box, see Figure 5.23(a).

(a)

(b)

Figure 5.22 Early electrification, New York, USA. (a) The Great Blizzard in Manhattan, 1888, demonstrating a plethora of overhead wires including telephone, telegraph, fire, burglar alarm and lighting systems, carrying DC and AC voltages ranging from 5 V to 6 kV. (b) Showing Manhattan, a post-card city lit-up, looking towards the East River, 1884.

Consequently, within the box, an urban railway system was required to convey passengers from main-line stations into the city, the first city transport system world-wide. This London urban railway system began construction in 1890.

These underground lines were initially built in London by adopting a simple cut-and-cover approach; however, this caused considerable surface level road and housing disruption. The cut-and-cover approach was superceded by tunnelling, made possible North of the Thames by the stiff blue clay of the London subsoil, which facilitated the use of a simple Victorian tunnelling technique, using a cylindrical steel shield, that could be forced forward through the clay in the tunnel direction, whilst navvies removed the clay from inside the shield face.

The new underground trains were initially hauled by steam tank engines. There were laughable suggestions at the time that locomotive exhaust cleared throats and

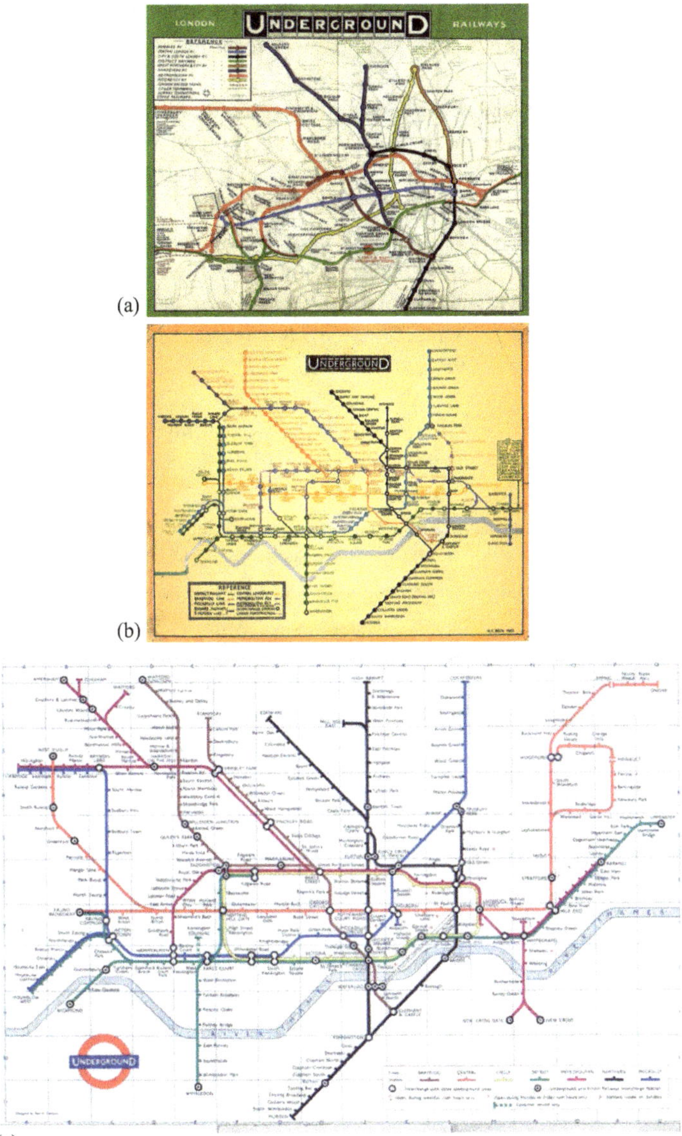

(a)

(b)

(c)

Figure 5.23 Expansion of the London Underground system, 1890–2000. (a) 1908, Early London Underground map, showing the city central box; (b) 1931, Later symbolic Underground map; (c) 1962, More recent symbolic Underground map, the central box is still visible in this schematic.

lungs; however, by the start of the twentieth century, such locomotives were relatively quickly fitted with condensing apparatus to allow them to recover exhaust steam reducing tunnel pollution.

The author can clearly remember another vignette, in the late 1950s, at Kings Cross Station seeing a train hauled by a tank locomotive with condensing apparatus, emerging from the Underground tunnel in clouds of steam and smoke, pounding up the bank into its separate platform.

These steam locomotives were soon superseded by more powerful, cleaner electric locomotives, taking power from ground-level electrified rails.

London's urban railway subsequently became named 'the Underground', as shown in system maps in Figure 5.23(a)–(c), now consisting of 24 lines.

The electric locomotives were in turn superseded by continuous, multiple-unit, electric train sets powered from the same electrified rails, using a four-rail DC system. The original electrical configuration was a nominal 630 V DC, with −210 V centre conductor rail and +420 V outside conductor rail and the running rails being used as an Earth connection.

This was the default configuration wherever deep-level London Tube trains operated. However, from 2016, to facilitate more powerful trains-sets, the sub-surface network has been reconfigured to use a 750 V DC configuration, −250 V and +500 V rails, and by 2019 almost, the entire sub-surface network now uses this configuration.

Refer to Duffy [16] for an excellent description of how the London Underground's electrical technology powered the trains but also enabled the signalling that controlled their tight-packed scheduling needed by a large urban system.

See also in Section 8.4.2 a most up to date, larger surface Urban Railway Network in Shanghai (上海), China.

5.6 Oil and gas supply

5.6.1 Oil

Figure 5.24 shows an example, in 1890, of another early modern-day oil resource extraction, in the then Russian Empire, now present-day Azerbaijan.

The photograph, Figure 5.24(a), shows multiple individual wells being drilled using large numbers of wooden oil derricks, with oil being transported, under well pressure, through short pipe-lengths to jetties, then onto vessels to be shipped elsewhere.

Compare this image with the start of modern-day commercial oil production, in the early 1860s, in Pennsylvania and New York states in the USA.

The graph, Figure 5.24(b), is also interesting as it shows the rise in oil extraction in Azerbaijan towards the end of the nineteenth century, followed by subsequent rises leading up to the First and Second World Wars.

Then, from 2000 to 2015, following the identification of new fields at Chiri, Azeri and Phase 3, including the use of new extraction technologies, a meteoric rise in output as the new fields came on-line, followed by a subsequent rapid collapse in the 2020s, as the field was exhausted.

(a)

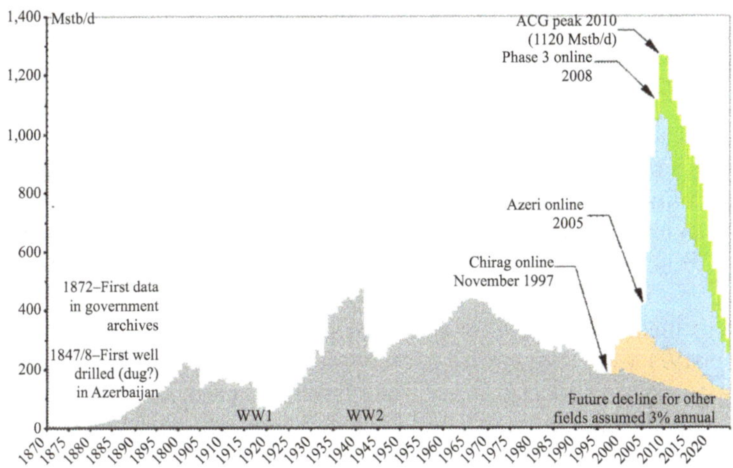

(b)

Figure 5.24 Development of Oil production in Azerbaijan, 1870–2022. (a) Post-card depicting the Bibi-Heybat deposit, Nobel Brothers oil field, Azerbaijan, 1890, Digital image Беиинг и Рааб *public domain courtesy Wikimedia Commons. (b) The gradual rise and fall of Azerbaijani oil production, creative commons attribution ShareAlike 3.0 License.*

5.6.2 Gas

The earliest records of gas supply in United Kingdom were when Thomas Shirley described an inflammable gas seeping from a coal mine in Wigan, 1667. In 1727, a gentleman named Carlisle Spedding in England heated his office with methane gas from a local coal mine.

The first commercialised natural gas, produced from coal, was extracted in Britain, 1785, and used to light streets and houses, while gas lighting went on public display in Paris, 1801.

It was a German inventor, Frederick Winzer, 1799, who pioneered the expansion of gas lighting in United Kingdom and France, with the idea to store gas, produced from a central gasworks, in large gasholders and deliver it to customers through a network of pipes.

Up until these innovations companies deploying gas lighting each used their own individual gasworks.

William Murdoch, an engineer from Ayr in Scotland, had walked from Ayr to Birmingham to work for Boulton and Watt. From his experiences heating coal to make coal-gas in Ayr, he investigated coal-gas distillation processes for illumination whilst making small-scale lighting experiments in his own house. He then worked for Boulton and Watt on similar experiments in 1792, but whilst there was ongoing patent litigation underway in Boulton and Watt's main steam engine business, the gas-lighting project was dropped.

However, Gregory Watt, James Watt's second son, while travelling in Europe saw Lebon's gas demonstrations and wrote to his brother, James Watt Jr., informing him of this potential competitor. This prompted James Watt Jr. to expand the gas-light experiments at Soho Works scaling up William Murdoch's technology, leading to a first commercial gas-light application.

After an initial installation at the Soho Works, 1803, the company prepared an apparatus for the textile firm, Philips and Lee, in Salford, Manchester, 1805, their only major gas equipment sale until 1808. George Augustus Lee, the mill's owner, was a motivating force behind the apparatus' development. He had an avid interest in technology, introducing a series of innovations at his Salford Mill, such as iron textile frame construction and steam heating and was encouraged by Boulton and Watt's development of this gas-light technology, Figures 5.25.

5.6.3 Early pipelines

In about 500 BCE, 2,000 years before Thomas Shirley, Chinese workers had started using crude bamboo pipelines to transport gas seeping from the Earth's surface, using that gas to heat and boil seawater, obtaining drinkable water.

By 1807, the first gas distribution pipeline was installed in London's Pall Mall, and in 1814, London had its first gas streetlights, the gas being delivered via up to 26 miles of underground piping from a gasworks similar to that shown in Figure 5.20(b).

Complex coal-gas works sprung up all over the cities of Western Europe and the United States, with expanding gas pipe networks to supply the coal-gas from works to housing.

In 1865, the first fully successful oil pipeline was laid, in Pennsylvania USA, see Edwin Drake, using wrought iron pipes with highly reinforced joints, transporting 1,950–2,000 barrels of oil daily across five miles of land.

(a)

(b)

Figure 5.25 *Early European gas production from coal. (a) Historic wrought iron manufacture, Coalbrookdale, United Kingdom, using coke manufactured from coal. (b) Drawing coal-fired retorts at the Great Gas Establishment, Brick Lane, London, Monthly Magazine, 1821.*

5.7 Summary

This chapter has demonstrated the development and growth of key energy production technologies over 130 years from the nineteenth into the early twentieth century, including the concept of the Energy Vector.

The steam Energy Vector came from the following fuels:

- Coal;
- Gas;
- Oil.

These facilitated:

- The first reciprocating engine prime movers using steam from coal are for:
    - Mine pumping;
    - Locomotion;
    - Ship propulsion;
    - Electrical power generation.
- The first reaction, impulse, Rateau and Brown-Curtis steam turbine prime movers using steam raised from coal are for:
    - Ship propulsion;
    - Electrical power generation.

The water Energy Vector then facilitated large hydro-electric dams using:

- Large slow-speed reaction turbines for lower head dams;
- Smaller higher speed Pelton wheel impulse turbines for higher head dams;
- All for electrical power generation.

The gas and oil vectors facilitated:

- Lighting;
- Power for gas and oil engines.

Together the consequences of these Energy Vectors have been the progressive development of:

- Steam reciprocating engine design;
- Steam turbine design;
- Water turbine design;
- The consequential widespread development of electrical energy power networks.

These Energy Vectors have facilitated the following transportation:

- Sea and river transport;
- Road transport;
- Rail transport.

The Energy Vectors have facilitated the acceleration of transportation as below:

- Movements of fuels;
- Movements of people.

Movements of goods.

Chapter 6

Modern age II, 1930–2000 CE, gas, heat engines and aviation

6.1 Introduction

This chapter considers the effects of current scientific knowledge on the conversion of energy from plants, water, hydrocarbons, gases and minerals from around the world.

This will also cover the increasing effectiveness and transportability of fossil fuels by land and sea, but their deleterious polluting effects and the influence of energy efficiency on their economy and pollution.

The importance of heat storage and flow in energy deployment and of networks, including electricity, for the sharing of power and the enablement of distributed energy sources.

The important part played by electricity, oil and gas networks in the distribution and effective application of energy sources and particularly of the way that is likely to affect the future development of renewable technologies and of energy supply.

Modern electrical and communication technologies are also facilitating energy flows for improved manufacturing, planning, economic efficiency and development.

Improved manufacturing techniques meant that we could construct engineering facilities at lower cost and in shorter times than hitherto.

The surprising abilities of new energy and communications technologies to mitigate the deleterious polluting effects of traditional energy sources.

The development of renewable technologies to meet energy needs without using fossil fuels and the essentially distributive nature of this new resource.

The possible influence of future Energy Vectors alongside electricity, for example hydrogen or other alternatives.

6.2 Modern power and energy units

Energy or work is measured in the following modern units:

- The kWh the physics unit is the Joule, 1000*Ws=1 Joule
- The MWh = 3.6 GJ = 1 tonne of oil equivalent (TOe)

Figure 6.1 *Drax, Britain's largest thermal power station. Generating 3,000 MW, burning biomass, imported from the USA, stored nearest camera, with some coal*

- The TWh = 1 MTOe
- 365 TWh = UK's approximate annual electrical energy consumption.

Power or the rate of doing work:

- 1 MW propulsion power of a Cambridge-London electric train;
- 3000 MW output of Drax Power Station, Yorkshire, United Kingdom, see Figure 6.1;
- 1 TW = 10^6 MW;
- World's installed electrical power capacity = 6.1 TW (2014)
- World's used electrical power output = 2.2 TW (2013)
- World total power production from all fuels including electricity = 19 TW
- P_e=Ratio of electrical /total power = 1/9, an economic development indicator
- P_e for a developed country, for example United Kingdom =1/3
- P_e for a developing country, for example Ethiopia =1/10

6.3 Modern petrol and oil power

6.3.1 Internal combustion engines

Petrol and diesel internal combustion engines had been developed at the end of the nineteenth century.

The fuel for petrol engines is derived from refined products of the complex oils produced by the oil-fields described in Chapter 5. The refining of oil had produced a wide variety of valuable commodities, including:

- Petrol
  Used to power road vehicles;
- Diesel oil
  Used to power agricultural tractors, heavy vehicles and smaller marine vessels;
- Jet fuel
  Used to power aircraft;
- Kerosene
  For lighting;
- Paraffin
  For home heating and lighting;
- Heavy fuel oil
  For large ships either steam- or diesel-powered;
- Liquefied petroleum gas
  Fuel for crop-drying, home heating and barbecues;
- Petrochemical feedstock
  For the manufacture of chemicals, synthetic rubber and various plastics.

The petrol engine, developed by several engineers, used the most energetic product from refined oil, which is beneficial for powerful performance.

However, the diesel engine, developed by Otto Diesel, uses the less-refined but cheaper fuel, because of that the diesel engine design incorporates some valuable ignition and thermodynamic benefits, which the petrol engine lacks.

6.3.1.1 Petrol engines

The modern four-stroke, Otto-cycle, petrol engine powers vehicles using electric spark ignition to explode the petrol fuel in the cylinder head.

This was the technology that has been used from the late nineteenth century to the present day but has been refined by many improvements.

The petrol engine ignition system and four-stroke cycle have proved more reliable than other engines, while the use of the petrol delivered a high energy output. So, petrol engines became the accepted power units for a wide variety of vehicles from small runabouts up to the largest luxury cars, over a period from 1900 until the late twentieth century.

The largest petrol engines were later also used to power fighter aircraft, for example the Spitfire, Figure 6.5(a), whose Rolls Royce V12 Merlin petrol engine with carburettors and spark ignition delivered an enormous power capacity.

In the late twentieth and early twenty first centuries here was a period of concern about the presence of lead in the exhausts of petrol engines because of the use of fuel lead additives to prevent knocking and improve combustion.

However, the performance of petrol engines and cost of its fuel have been heavily challenged by the diesel engine in the latter years of the twentieth century.

6.3.1.2 Diesel engines

The diesel is a compression ignition engine of enormous potential, conceived by Otto Diesel in Germany as an internal combustion engine like the petrol engine but without spark plugs.

Fuel is injected in accurate quantities into each cylinder and ignition is achieved by the compression and consequent heating of the fuel under high pressure.

The cycles of petrol and diesel engines are useful and are compared in Figure 6.3.

For a petrol engine the piston compresses the injected petrol–air mixture up to the top-dead centre (TDC) when the mixture is ignited by a spark, Figure 6.3(a).

However, for a diesel engine the diesel–air mixture is compressed further to TDC, causing the fuel–air mixture to heat rapidly and spontaneously ignite, Figure 6.3(b), this is assisted by arrangements to swirl the fuel when it ignites.

The petrol and diesel operating cycles are compared in Figure 6.2.

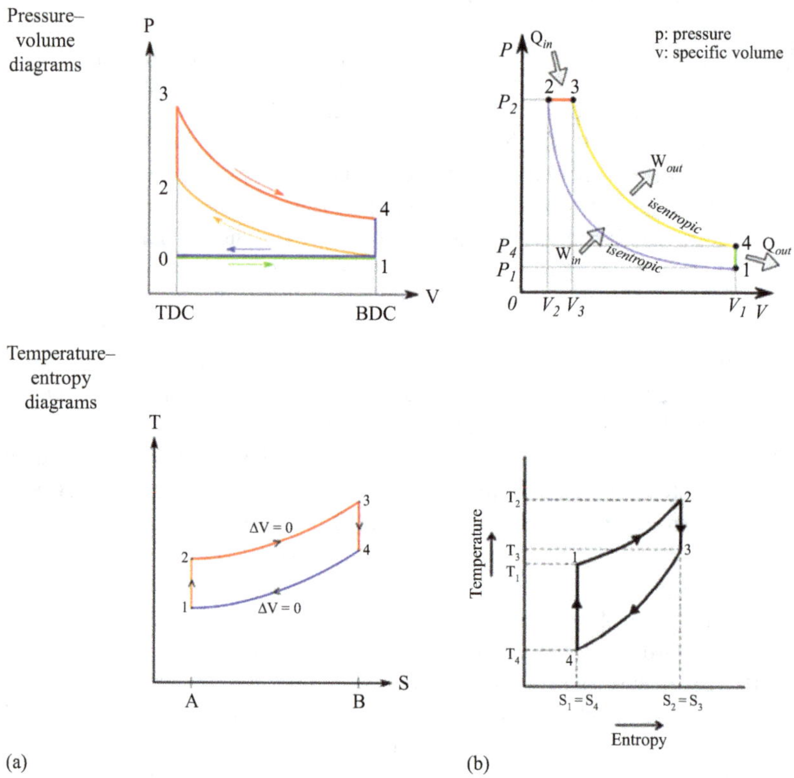

Figure 6.2 Comparison of internal combustion engines, petrol or Otto cycle and diesel cycle. Pressure-volume diagrams and Temperature-entropy diagrams of (a) petrol engine and (b) diesel engine.

The operational advantages of diesel engines are as follows:

- Diesel fuel is cheaper than petrol, because the latter is more refined;
- Otto Diesel's compression ignition has proved more reliable than spark ignition;
- Modern diesel combustion chamber designs, with higher cylinder pressures and fuel flame swirl, have proved more reliable than spark ignition systems;
- The net result is that a diesel engine releases more energy from the fuel per cycle than a petrol engine.

The operational disadvantages of the diesel engine are as follows:

- Diesel engine exhaust is more toxic than that of petrol engines;
- Maintenance of high-pressure diesel injection systems is alleged to be more challenging than for spark ignition engines. Although modern manufacturing techniques of diesel engines have largely eliminated this disadvantage.

Whilst diesel engine's earliest applications were for motor transport, a breakthrough occurred when the Imperial German Navy selected Diesel's technology to power their first submarines, before the First World War, see Figure 6.5(a) and (b).

Other countries, including the Royal Navy, followed suit and this naval application of diesel engines played a large part in their later modern industrial development.

Diesel fuel contains more energy than the same volume of petrol. The difference can be quite large, it is not unusual for a diesel car engine to have an official average 70 mpg figure, compared with about 50 mpg for an equivalent petrol model. However, recent environmental concerns have placed restrictions on the use of diesel fuel. There are, however, a growing number of diesel fuel alternatives including renewable vegetable oils, such as palm, soybean, sunflower, peanut and olive oil.

Diesel technology now powers almost all land transport heavy vehicles from tractors, see Figure 6.5(c) and (d), to trucks and railway locomotives, see Figure 6.6(a).

At sea diesel technology is used for small fast launches, through to warships up to enormous 400,000 tonne container ships, see Figure 6.6(b) and (c).

In the twentieth century, the applications of diesel engines rapidly increased, and the diesel engine has provided heavy lifting power for modern economies around the World, demonstrated in the figures below.

6.3.2 Pipelines

In Canada and the USA in the nineteenth century, oil was initially conveyed by barrel from well-heads to refineries, but then by short small-bore pics to rail heads and then by rail tank wagons, refer to Edwin Drake, c1860s.

Long distance oil and petroleum pipelines commenced building in North America in the 1930s to 1940s, see Figure 6.3, in response to the needs of the Second World War.

Figure 6.3 Big and Little Inch petroleum pipelines from Texas to New Jersey, built 1942–44, USA

Similarly, in response to the Second World War, oil pipelines started to be developed a large-scale in the United Kingdom also.

Figure 6.4(a) shows the pipeline system sending oil to the main RAF and USAAF air-fields in South and Central England, whilst Figure 6.4(b) shows the pipes being laid.

However, these pipelines were a faint replica of the complexity of the pipelines being installed in the USA and Canada.

A demonstration of the effect of war on the development of technology was the provision of a pipeline to supply oil and other fuels to the armies, through flexible pipes across the Channel from England to France, for the Allied re-invasion of the European continent, in 1944. The system had the code-name PLUTO, Figure 6.5. Early applications of the diesel engine were first to submarine propulsion and then to tractors for agricultural use, as shown in Figure 6.6.

An interesting comparison can be made between the application of petrol- and diesel-powered piston engine propulsion with WWII fighter aircraft.

Whilst Britain developed the Spitfire, with a petrol-fuelled, carburettor fitted, Rolls Royce Merlin engine.

Germany on the other hand developed the Me 109 fighter with a diesel-fuelled, injected, Daimler-Benz DB 605 engine.

RAF pilots in the earlier stage of Second World War noticed that, while the Spitfire had superior manoeuvrability than the Me 109, the Spitfire had some shortcomings because of issues with the engine's fuel supply.

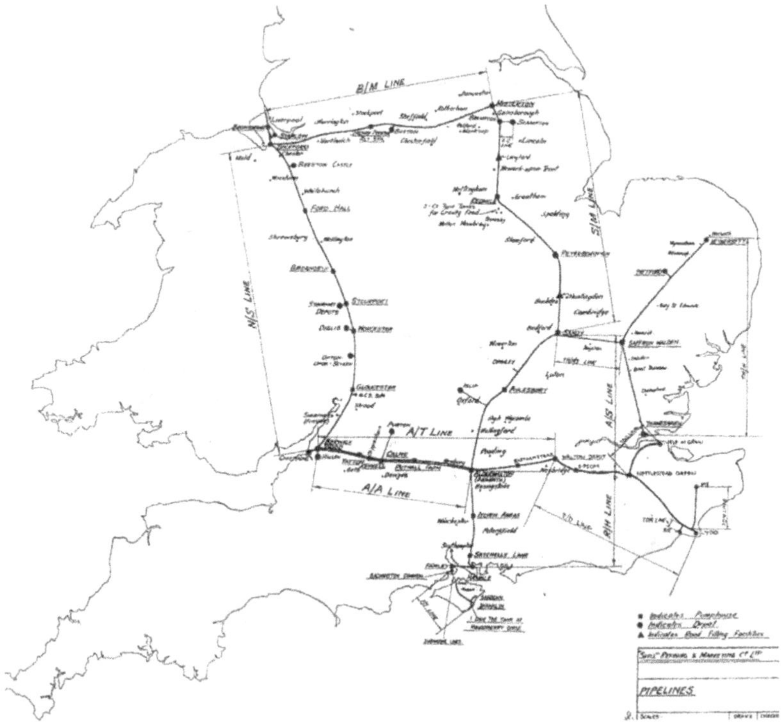

(a)

(b)

Figure 6.4 Oil pipe-line network and laying, 1944, UK: (a) oil pipe-line network; (b) laying steel oil pipe-lines

(a)

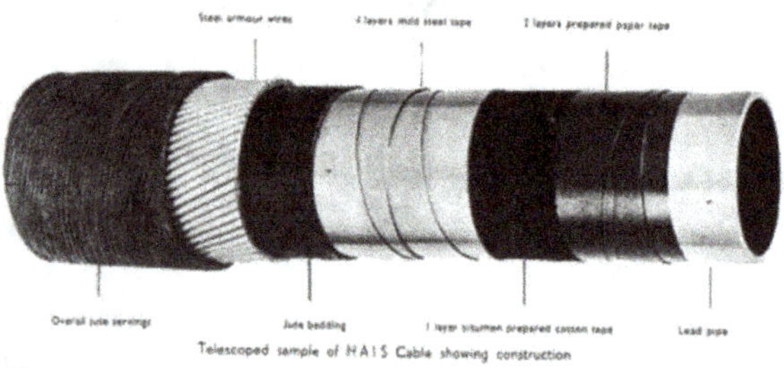

(b)

(c)

Figure 6.5 The PLUTO oil pipeline to supply the Army for the re-invasion of Europe, Second World War: (a) map of oil pipe-line network; (b) a section of the flexible PLUTO oil pipe; (c) a pipeline reel of the being laid, across the English Channel, 1944

Figure 6.6 *Early applications of diesel engines, 1906–68: (a) Imperial German Navy U 1 Submarine, 1906; (b) A submarine diesel engine, 1906; (c) Diesel farm tractor, 1941; (d) Diesel farm tractor, 1968*

(a) (b) (c)

Figure 6.7 Examples of large-scale diesel applications: (a) Trains: Deltic 55 022 Royal Scots Grey, 2009 W 100 tonnes, capable of 100 mph; (b) Ships: Wärtsilä RT-flex96C, 2006 turbo-charged, 2-stroke, low speed diesel engine L 27 m, H 13.5 m, W 2,300 tonnes; (c) Container ship, EMMA MAERSK, W 171,542 tonnes, diesel-powered by a Wärtsilä RT-flex96C engine, 2006

The diesel engine was then developed for railway locomotives and ship propulsion applications, see Figure 6.7.

6.4 Modern gas power

6.4.1 Gas lighting

In the late eighteenth and early nineteenth centuries the production of iron had increased, and the development of steel production commenced.

Steel manufacture requires carbon and that is supplied by reducing coal to coke in heated retorts. A natural by-product of coke manufacture is coal-gas.

Increasing coal production in the nineteenth and twentieth century lead to a more-or-less continuous supply of coal-gas, produced when coal was heated in retorts to make coke for steel production, see Figure 6.8(a). This coal-gas was initially used for lighting, particularly in Europe, becoming an important industrial consequence of the coal and steel industries.

We saw in Chapter 4 how in the USA and Canada the demand for lighting was initially met primarily by the exploitation of oil. However, in other parts of the World, lighting was being provided by coal gas.

Between 1870 and 1905 the number of beehive coking ovens in the US increased from approximately 200 to nearly 31,000, producing nearly 18,000,000 tons of coke a year in the Pittsburgh area alone.

Modern natural gas industries have expanded since the 1970s, from gas extracted from oil and gas reservoirs largely located in the North Sea but with some isolated onshore on the mainland of Great Britain, Figure 6.8.

(a)

(a)

(b)

Figure 6.8 Coal to coke for manufacturing steel, coal-gas is a by-product of coke manufacture: (a) last coking plant supply to Pittsburgh steelworks, USA; (b) modern coke oven battery supplying steelworks and coal-gas works, Janjul, India; (c) modern plant making coal–gas from coke, Seattle, USA

6.4.2 Gas engines

A gas engine is an internal combustion engine running on gaseous fuel, in the nineteenth century this was originally from coal-gas but progressively became gas from oil refineries and by the twentieth century gas from land-fill sites also became used.

These gas engines started to be introduced into factories in the 1880s to provide power without the need to raise steam using a large and costly boiler.

Gas engines became widely used in the nineteenth century in many industrial processes, spinning, weaving, paper mills and such applications lead to the use of wider ranges of fuels, for example producer gas, biogas, landfill or natural gas.

In the UK and English-speaking countries, the term is unambiguous. In the United States, due to the widespread use of 'gas' as an abbreviation for gasoline or petrol, such an engine might also be called a gaseous fuelled or natural gas engine or spark-ignited.

In modern usage, the term gas engine refers to a heavy-duty industrial engine capable of running continuously at full-load for periods approaching a high fraction of 8,760 hours per year, unlike a gasoline automobile engine, which is lightweight, high-revving and typically runs for no more than 4,000 hours in its entire life.

Typical gas engine power ranges from 2 kW to 5 MW such as are shown in Figure 6.9.

6.4.3 Private homes

The heating of private homes can be compared with the energy used to grow trees of a similar size.

An example is given in Figure 6.10, which may surprise readers, that our addiction to electricity and the use of fossil fuels to heat a modest home is trumped by the extraction of energy from the sun to grow a tree of similar size.

6.4.4 Early gas turbines, lessons from steam

Charles Parsons had developed the first steam reaction turbine in the late 1880s, but at that time he also worked on an experimental high-speed torpedo, registering a patent in 1894.

For the torpedo's propulsion he used a small version of his reaction steam turbine but powered it by gas produced from an explosive mixture. This early reaction gas turbine (GT) design had many problems, his conclusion at the time was that maybe a GT would be impracticable.

However, a first successful GT, see Figure 6.11, was built in 1903 by Jens William Ægidius Elling, a Norwegian pioneer. The machine consisted of a three-cylinder, multistage reciprocating compressor, a combustion chamber and turbine to produce an estimated 8 kW output power.

It operated as follows, air was supplied by a compressor, mixed with liquid fuel and burnt in the combustion chamber, then exhausted through the impulse turbine. Elling's GT did not find an immediate propulsion application and was perhaps superseded by petrol and diesel driven engines, but it laid the foundation for the future of GTs and their later entry into aviation propulsion.

Figure 6.9 *Reciprocating gas engines, smooth base load power: (a) Early Crossley gas engine, 2.2 kW, 1880; (b) Crossley gas engine, 112 kW, 1915; (c) GE Jenbacher gas engine, 5MW, 2010*

(a) (b)

Figure 6.10 Comparison between the energy used to heat a small family home and that needed to grow an oak tree of similar size: (a) small UK house, electricity use 5 kWh/day, gas use 22 kWh/day Source: Cambridge House; (b) 100 year old Oak tree, 12 m dia Insolation 900 kWh/day Tree photosynthesises 57 kWh/day from average N European solar radiation Source: Prof Dr Keller, University of Siegen, Germany, advice from Dr M Joynes, MIET

6.5 Aviation

6.5.1 Aviation begins, effects of war

Practical aviation started with the Wright brothers in 1904 but earlier experiments had paved the way. Early flight was achieved with fixed-wing biplane structures.

Aviation took huge steps forward during the World Wars of the twentieth century.

To demonstrate this, in the 1914–18 First World War the technology was initially overwhelmingly fixed wing biplane aircraft, predominantly powered by a range of petrol fuelled in-line or rotary reciprocating engines. The following lists demonstrate the number of military aircraft manufactured and lost per combatant countries, as follows:

- France: 67,987 aircraft produced, 52,640 lost;
- British Empire: 58,144 aircraft produced, 37,973 lost;
- German Empire: 48,537 aircraft produced, 27,637 lost;

Figure 6.11 Ægidius Elling's first gas turbine engine, 1903

- Kingdom of Italy, 20,000 aircraft produced, losses unknown;
- United States of America, 15,000 aircraft produced, losses unknown;
- Austro-Hungarian Empire, 5,431 aircraft produced, losses unknown;
- Russian Empire, 4,700 aircraft produced, losses unknown;

  First World War Total: 219,799 aircraft produced.
  At least 118,250 aircraft lost.

This production and operational experience provided a huge stimulus to post-war civil aviation, which developed during the inter-war years.

The Second World War, which some have described as a mere extension of the First World War, maybe said to have been resolved by aviation.

Again, to demonstrate this War's scale from 1939 to 1945, the following lists the number of military aircraft manufactured and lost per combatant country, overwhelmingly of fixed wing monoplane design, powered by a range of in-line and rotary reciprocating engines petrol or diesel fuelled with latterly some jet engines, were as follows:

- United States of America: 324,750 aircraft produced, 95,000 lost;
- Soviet Union: 157,261 aircraft produced, 90,110 lost;
- British Empire: 131,549 aircraft produced, 37,973 lost;
- Nazi Germany: 119,907 aircraft produced, 40,000 lost;
- Japan: 76,320 aircraft produced, 35–50,000 lost;
- Kingdom of Italy: 11,122 aircraft produced, 5,272 lost.

- **Second World War Total: 820,909 aircraft produced, four times that of First World War.**
- **At least 264,455 aircraft lost.**

The aviation lessons learnt in the Second War laid the foundations for civil aviation today. These lessons were from experience operating all kinds of propeller-driven aircraft, in particular the logistics of aircraft operations, communications, radar, safety and above all air crew and ground crew training.

An interesting sideline to the successful development by Rolls Royce of the Merlin petrol engine was that the fuel was supplied to the engine via carburettors, Figure 6.12(b), with the engine being fitted to British Spitfire and Hurricane fighter aircraft.

(a)

(b)

(c)

Figure 6.12 Engine improvements under war-time conditions, 1940: (a) gun camera view of a Spitfire firing into a German He 111 bomber; (b) Spitfire's Rolls-Royce Merlin 12-cylinder in-line engine, carburettor cuts-out under negative G manoeuvres; (c) carburettor problem resolved by RAE engineer, Beatrice Shilling

However, in combat RAF pilots reported, that when manoeuvring close to an enemy in a dogfight, Figure 6.12(a), engine cut-out problems could occur due to carburettor starvation. This is clearly audible in the opening scenes of the film '*The Battle of Britain*', as a Hurricane fighter performs a victory roll, 1969'.

The problem was solved by a Royal Aircraft Establishment engineer and motor-cycle enthusiast, Beatrice Shilling, Figure 6.12(c). This resulted in the fitting of the RAE Restrictor, developed by her, to the carburettors of all RAF fighter aircraft, solving the problem. That modification was remembered by RAF pilots affectionally as '*Miss Shilling's orifice*'.

(a)

(b)

Figure 6.13 Sir Frank Whittle and his pioneering radial jet powered aircraft: (a) Frank Whittle, his pioneering aircraft gas turbine jet engine W-1b, thrust 3.8 kN 1940; (b) the pioneering Gloster E.28/39 aircraft powered by Whittle's engine, 1942

6.5.2 Aviation and jet power

Between the World Wars, following Elling's GT prototype, there was an intense period of GT engine principles developments, because of the engine's potential power and the realisation that fixed wing aircraft speeds needed to increase.

This culminated early in Second World War, in 1940 with the development by Frank Whittle of his GT, jet aero engine, Figure 6.13, fitted into the first jet fighter aircraft in 1942. Germany also developed a GT propelled aircraft.

Whittle's engine was simple and compact with a single-stage, centrifugal compressor, 10 combustion chambers and a single-stage axial impulse turbine on a common shaft with the compressor, rotating at up to 16,000 rev/min. The purpose of the engine was to provide thrust to the aircraft, so its performance was measured and early engines delivered up to 8,000 N thrust, sufficient to get the aircraft shown in Figure 6.13 up to 690 km/h at a height of 4,600 m.

There were many other subsequent innovations and imitators.

6.5.3 Gas turbine advances

6.5.3.1 Improving thrust and efficiency

Figure 6.14 shows the evolution of plant efficiencies for GTs compared with other background technologies, including:

- Steam

  Where efficiency is limited by the highest achievable steam temperature, the highest stresses which affordable steel materials can attain, the lowest temperature at which steam can be condensed. These effects achieved using simple, regenerative feed heating or single or double reheat steam plants;

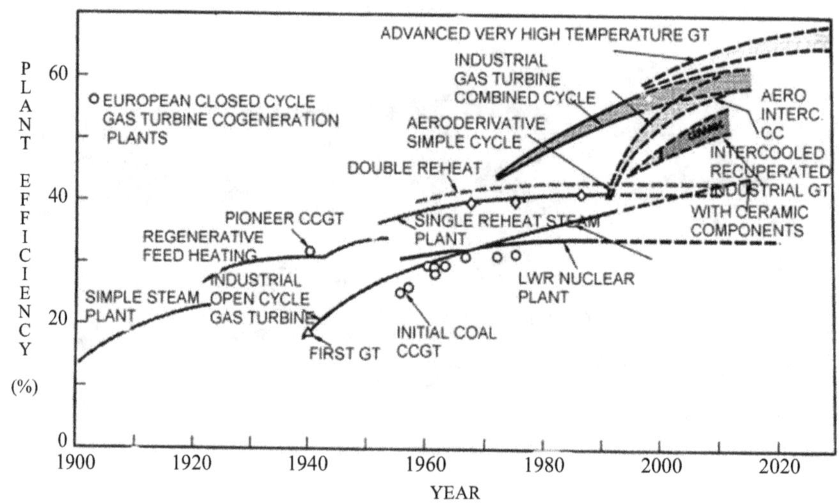

Figure 6.14 Time evolution of steam and gas turbine efficiencies, comparing technologies, McDonald, 1994

- Nuclear, light water reactor technology;
- GT;
- Combined cycle gas turbines (CCGTs).

6.5.3.2 Fighter aircraft, thrust and endurance

Many jet aircraft followed Whittle's prototype, Germany was particularly energetic in these developments, with many new innovations as shown in Figure 6.15.

6.5.3.3 Passenger aircraft, power and endurance versus efficiency

Figure 6.16 shows how the revenue per kilometre (RPK) of passenger aircraft has improved progressively since 1950. These improvements have been achieved by improved aircraft structural design but most prominently by the improvement of airframe structures, surfaces and power control.

Table 6.1 shows a 28-year period of improving efficiencies for a variety of competitive Airbus and Boeing jet passenger aircraft.

It can be seen from the Table that Airbus and Boeing aircraft efficiencies have improved, as follows:

- 1992–98: 3.49–3.76 L of fuel/seat/km;
- 2017–20: 2.42–2.48 L of fuel/seat/km.

These improvements were due to changes in aircraft wing and frame design combined with improved engine thrust and efficiency.

6.6 Power generation and aviation

6.6.1 Turbines, aero and power technologies

Aviation has had a profound effect on the development of heat engines, whether for aircraft, electricity generation or ground propulsion.

On the one hand aircraft needed very lightweight engines but with enormous thrust; however, on the other they for electricity generation or ground propulsion they must have reliability and a long life.

As a result, light-weight, long-lived and reliable jet engines have quickly superseded the previous heavy rotary or inline reciprocating aero engines, perfected during the First and Second World Wars.

Figure 6.17 demonstrates aero engine evolution, from a UK perspective.

An important outcome from the aviation industry for jet GT engine manufacture was a better understanding of the importance of blade shape and strength to improve engine outputs, see Figure 6.8(c) and (d).

Other important engineering lessons, learnt from both reciprocating and GT aero engines, are as follows:

- The application of high component material standards;
- The need for better component dimensional standards;

(a)

(b)

(c)

Figure 6.15 German jet aircraft developments during Second World War: (a) Heinkel He 178, world's first aircraft to fly with a single turbo-jet prototype; (b) Messerschmitt Me 262, operational twin turbo-jet fighter; (c) Horten Ho 229, captured prototype twin turbo-jet fighter/bomber

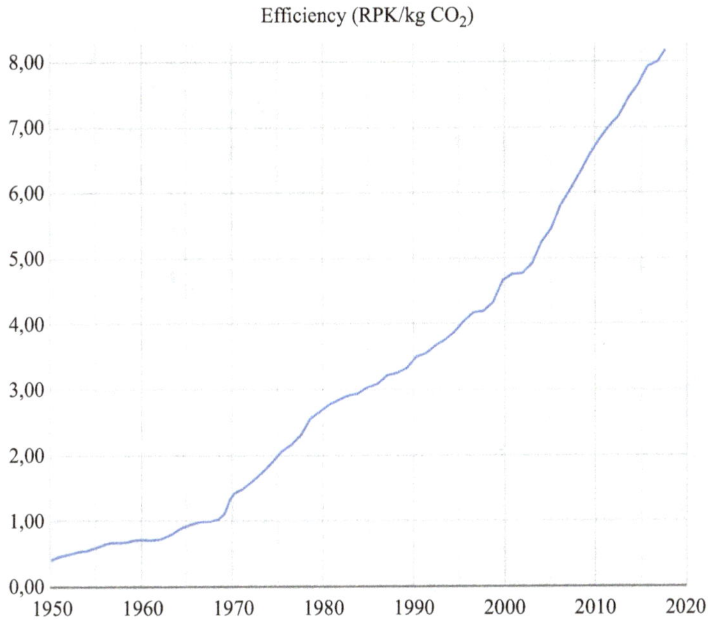

Figure 6.16 Passenger aircraft efficiencies improve from 0.4 to 8.2 from 1950 to 2018 in revenue per km RPK/kg CO_2 emitted per passenger

Table 6.1 Comparison of passenger jet aircraft fuel efficiencies, 1992–2020

Model	First flight	Seats	Range (km)	Fuel burn (kg/km)	Fuel per seat (L/100 km)
Airbus A340-300	1992	262	11,000	7.32	3.49
Airbus A380	2005	544	11,000	13.78	3.16
Airbus A350-900	2013	318	10,249	6.52	2.56
Airbus A350-900	2013	315	12,116	7.07	2.81
Airbus A350-1000	2016	367	10,243	7.58	2.58
Airbus A330-900	2017	300	8,610	5.94	2.48
Boeing 747-400	1988	393	10,192	11.82	3.76
Boeing 777-200ER	1996	304	10,251	7.57	3.11
Boeing 777-200LR	2005	291	9,208	7.57	3.25
Boeing 747-8	2011	467	11,000	10.54	2.82
Boeing 787-10 Trent	2017	337	10,240	6.24	2.31
Boeing 777-9X	2020	395	13,300	7.69	2.42

- Manufacturing standardisation and repeatability;
- Precision and best practices in manufacturing process.

The reader will recall early work, in the 1890s, on reaction-type steam turbine blading by Parsons, which were made from simple sprung steel sheet with short, almost flat blade shapes.

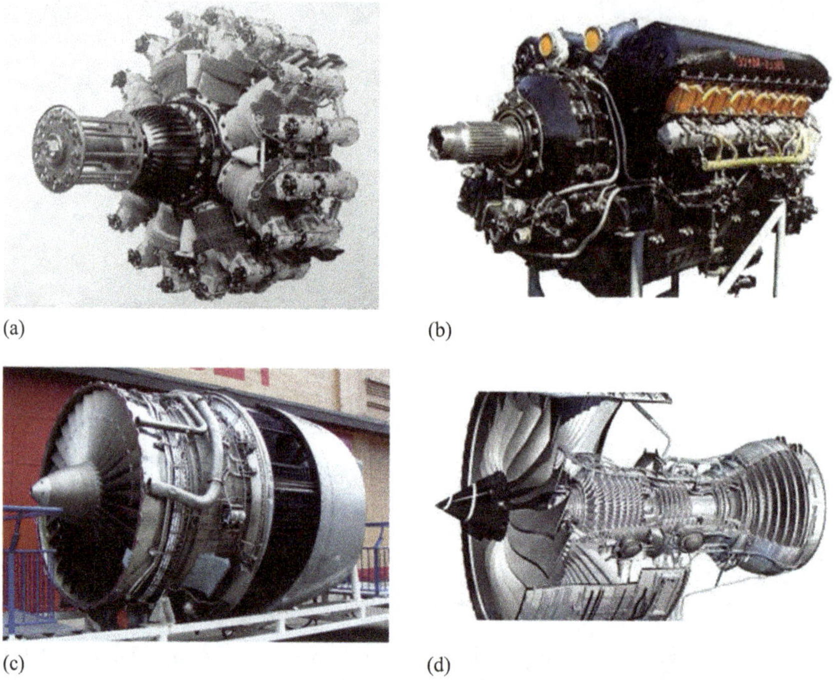

Figure 6.17 Evolution of aviation engines: (a) Bristol Hydra, 16-cylinder radial, 0.34 kW/kg, 1935; (b) Rolls Royce Merlin, 12-cylinder in-line, 0.38 kW/kg, 1939; reciprocating engines; (c) Rolls Royce RB 211, 3 shaft turbo-fan, 5.09 kW/kg, 1970; (d) Rolls Royce Trent, 3 shaft turbo-fan, 8.44 kW/kg, 1990; gas turbine engines

Subsequently, steam turbines evolved in the early 1900s, from the reaction type to Gustaf de Laval's impulse, Auguste Rateau's pressure-compounded impulse and Brown-Curtis' impulse turbine types.

Modern steam and GT blades use much more complex alloy materials and more complex shapes to achieve their improved output performance and efficiency.

Thus, new lessons have been learnt from steam and aviation turbine blades, about shapes, strength of materials and allowable temperatures.

6.6.2 Combined cycle plant development

Steam and GT blade materials and shapes have gradually been optimised, with lessons learnt from aviation GT engines and power generation gas and steam turbine prime movers.

This trickle-down has also been facilitated, in part, because power industry manufacturers were merging with other international businesses manufacturing aviation engines, for example Siemens, Rolls Royce and GE.

Figure 6.18 shows a modern CCGT power station in the USA. The station's gas and steam thermal cycles are shown in Figure 6.18(b).

The Brayton cycle describes the thermodynamic operation of the leading GT, while the Rankine cycle describes the thermodynamic operation of the following steam turbine.

The plant gas, steam and water temperatures are shown in Figure 6.18(c).

These figures demonstrate the benefits accrued from the absorption of material, manufacturing and thermodynamic lessons learnt from both power and aviation.

Figure 6.18 is an example of the most efficient thermal power generation achievable today, it is still fossil-fuelled but its performance is way ahead of the very best plants described in Chapter 4.

These benefits became the foundation for the development from 1980 to 2020 of CCGT power to achieve high efficiencies and long life. Initially this technology was developed for exploiting offshore oil and gas fields but was then deployed on-shore, becoming a significant factor in electrical power generation in Europe, USA and the Far East.

This is the type of fossil-fuelled thermal plant that could enable us to continue to provide electricity efficiently while we transition from fossil fuels to newer low carbon and zero carbon generators.

6.6.3 Heat engine comparisons

Table 6.2 provides a comparison of heat engine performance from the seventeenth century to the present-day, summarising developments in fuel, usage and efficiency for a wide range of power technologies.

Table 6.1 shows early, gradual heat engine improvements, their different applications and effectiveness and their progression from coal to coal-gas, to oil, to natural gas.

This heat engine comparison emphasises the importance to aviation of the development of lighter GTs with more efficient conversion of fuel to thrust and those GTs subsequently incorporated into power generation applications.

Consequent improvements in turbine blade and heat management technology have culminated in the development of combined heat and power plants generating electricity, with a GT coupled to a steam turbine, utilising the GT's hot exhaust for steam raising, driving a single electric generator, achieving overall efficiencies of 60%–65%, as shown in Figure 6.8.

Figure 6.8 and Table 6.1 taken together demonstrate the very substantial improvements made in the thermodynamics of energy conversion of heat engines over the last 50 years, with an important contribution from aviation over the past 100 years.

However, whilst these energy system efficiency improvements give cause for hope, they have also led inevitably towards increased energy consumption from fossil fuel usage, with further potential environmental damage.

Higher efficiencies are essential for reducing environmental impact but if higher efficiencies simply lead to increased energy usage from high carbon sources, our technologies are taking us in the wrong direction.

(a)

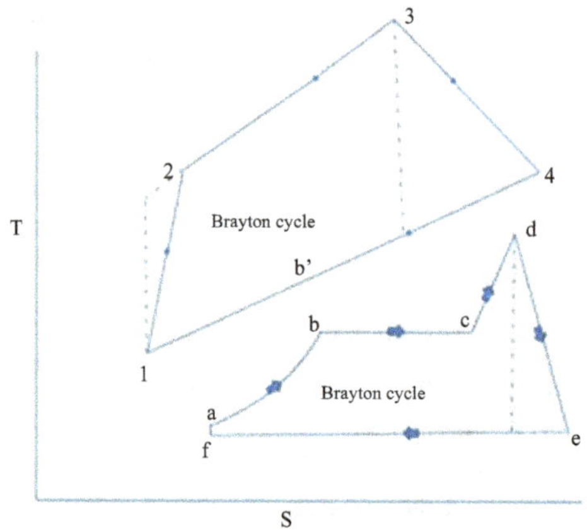

(b)

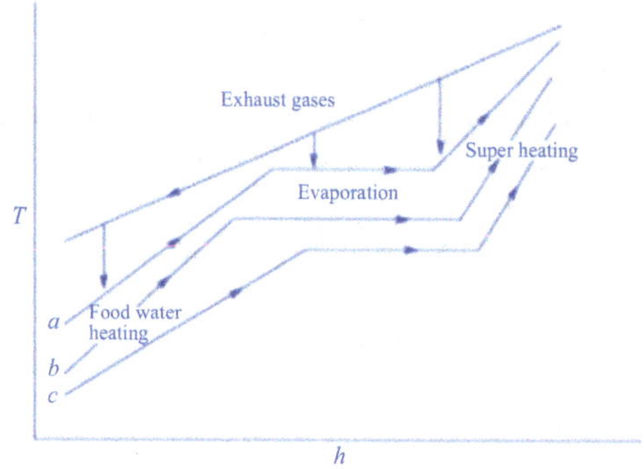

(c)

Figure 6.18 530 MW Combined Cycle Gas Turbine (CCGT) Power Station, California USA: (a) A CCGT power station, turbine units on the left, condenser on the right; (b) Brayton gas and Rankine steam thermal cycles, temperature, T versus entropy, S; (c) heat transfer from exhaust gases to steam, temperature, T versus enthalpy, h

Table 6.2 Heat Engine Efficiency Comparisons, 1698–2020

Year	Engine originator	Fuel	Engine type	Operating fluid	Control	Usage	Efficiency (%)	Power range
1698	Savery	Coal	Vacuum	Steam	Hand-operated valve	Water pumping	<1	<1kW
1712	Newcomen	Coal	Reciprocating single cylinder	Steam	Hand-operated valves	Water pumping	<1	2–10 kW
1795	Boulton and Watt	Coal	Reciprocating	Steam	Automatically operating valves	Water pumping	6	10–100 kW
1884	Parsons	Coal or oil	Reaction turbine	Steam	Governor operated valves	Electricity generation	7	2 kW–2 MW
1899	Ship prcpulsion	8	2 kW–20 MW					
1860	Otto and Langen	Coal-gas	Reciprocating	Gas	Throttle regulator	Vehicles	9	0.19–2.24 kW
1897	Diesel	Diesel Oil	2-Stroke Reciprocating	Gas	Throttle regulator and injection nozzles	Electricity generation	12	1–2 MW
1912	Ship propulsion	18	10.20 MW					
1930	Locomotives and Vehicles	20	1–10 MW					
1941	Whittle	Paraffin	1 Stage Impulse Turbine	Gas	Throttle regulator and injection nozzles	Aircraft propulsion	1.5 Thrust to weight	500 kW
1985	C A Parsons Ltd and CEGB	Coal	6 Impulse and reaction turbines	Steam	Governor operated valves	Large power station electricity generation	40	3300 MW
2020	Siemens, GE, ABB, Mitsubishi	Gas	Combined cycle gas and steam turbine	Gas	Impulse turbine	Electricity generation	65	100–750 MW
Steam	Impulse and reaction turbines							

6.7 Summary

Growing human ingenuity can be assessed by an increase in knowledge.

There were 5,300 years between the production of YangShao pottery and the development of Boulton and Watt's steam engine.

There were 4,942 years between the production of the first flint axes from Grimes Graves and Whittle's development of the jet-propelled aircraft.

These products, all the result of human innovation, ingenuity and the environment, in different areas of imagination and widely different parts of the World, demonstrate humans' artistic, scientific and engineering progress.

This chapter has shown the development of the last 100 years of the following modern energy production and transmission technologies:

- Fuels
  - Gas-from-coal;
  - Gas-from-oil;
  - The increasing significance of gas wells;
  - Exploitation of gas engines;
  - Development of GTs:
    - First for aviation;
    - Then for electrical power production.
- Electric power
  - The start, with slow-speed reciprocating steam engine driving generators;
  - High-speed steam turbines take over;
  - High-speed GTs take over;
  - Combined cycle high-speed gas and steam power cycles developed;
  - The development of the transformer, transmission equipment and controls needed to deliver that electrical power.
- Aviation
  - The inception of human aviation;
  - The influence on aviation growth of the two World Wars;
  - Improvement of aviation performance based on increasing engine power;
  - The importance of a high power to weight ratio to aviation engines;
  - The transition from reciprocating to GT engines for aviation power;
  - Higher efficiency aviation gas turbine engines;

  Methods to reduce aviation carbon footprint.

Chapter 7
Modern age III, 1900–2000 CE, electricity and networks

7.1 Introduction

This chapter will describe the influence of electricity on energy distribution and the part that it has played in the human World's large-scale industrialisation. The following nomenclature for AC system voltages will be used: low voltage (LV), 100–1000 V; medium voltage (MV), 1–100 kV; high voltage (HV), 100–345 kV; extra high voltage (EHV), 345–765 kV. It will demonstrate the importance that electrical networks play in distributing energy, first in cities, then through HV networks across countries and finally through EHV networks across continents.

In the future, electricity is likely to be of great importance as a carbon emission reducing technology. Although electricity can be costly to generate, it is easy to transmit over large distances and facilitates easy energy transmission. As renewable energy becomes more accepted as a supplier of clean energy, for example from:

- Dams;
- Solar panels;
- Wind wave or tidal resource electrical generator prime movers.

Then it will be likely that electricity will play an increasing part in reducing the industrial and commercial impact of fossil fuels.

However, an important present issue, arising more recently from electricity network developments and present-day economics, Stern [19], is that electricity generation, transmission and distribution costs are rising, leading to an important conundrum for future clean energy:

- Capitally intensive top-down energy networks, developed from original electricity development concepts;
- Versus, less expensive bottom-up energy network concepts with future potential benefits in developing countries.

This chapter will deal with the development from 1900 to 2000 of present-day top-down electrical energy networks and hint at possible future benefits from a bottom-up energy network approach.

7.2 Electrical power stations and networks

7.2.1 Reciprocating prime movers

The earliest electrical generator prime movers from 1880 to 1900 were steam reciprocating engines of 1–5 MW but from 1900 to 1910 were being overtaken by the high-speed steam turbine, with unit power ratings rising from 10 MW by the start of the twentieth century and up to 1,200 MW at the end of the in twentieth century.

After 1880–1910, the steam reciprocating engine continued for pumping water and other fluids.

However, diesel- or gas-powered reciprocating engines, of 4–20 cylinders, have continued for small-scale electricity production, of 100 kW–10 MW, in local power stations.

7.2.2 Turbine prime movers

The high-speed steam turbine developed rapidly at the end of the nineteenth century with unit power ratings rising from 1 MW at the end of the century, increasing to 1,200 MW per generator unit by the end of the twentieth century.

Low-speed hydraulic turbines have been employed in the hydro-electric generation industry, with outputs rising from 1 MW per generator unit at the end of the nineteenth century to 750 MW per generator unit at Three Gorges Dam, by the end of the twentieth century.

7.2.3 Prime movers, generators and power stations

Power station sizes have progressively increased as steam boiler, prime mover and generator technology advanced.

By the end of the nineteenth century, thermal power station output powers of 10–20 MW were becoming commonplace, with the USA taking the lead in large ratings.

Thermal power stations have steadily increased in size and rating throughout the twentieth century with generator sizes of 660–1,400 MW becoming commonplace and thermal power station peak outputs ranging for example from 3,300 MW in United Kingdom with China taking the lead with the largest power station rating 6,700 MW, Figure 9.3.

Hydro power stations have also steadily increased in size and rating to the current following huge schemes, whose potential peak outputs are shown in the following:

- Three Gorges Power Station on Yangtze River (長江), China, 22,500 MW;
- Itaipu Dam on Paraná River, Paraguay and Brazil, 14,000 MW;
- Guri Dam on Caroni River, Venezuela, 10,235 MW;
- Grand Coulee Dam on Columbia River, USA, 6,809 MW;
- Krasnoyarsk Dam on Yenisei River, Russia, 6,000 MW.

However, on a smaller scale, there has also been a steadily growing application of relatively small power generating units, exploiting industrial or locally available fuels for supplying electricity to local industrial hubs. This has been an important growth area, particularly in Northern Europe, where operators have been eager to exploit new customer bases and newly available energy sources, in terms of unusual fuels, such as agricultural biomass, pre-used cooking oils and the output gases from industrial plants, see for example Figure 9.6.

7.2.4 Electrical machines, increasing power-weight ratios and applications

The development of the electrical motor has been more radical than for the electrical generator because motor applications have been over a much wider industrial field.

Figure 7.1 gives an idea of the ranges of machine power/weight values now being deployed, whilst Table 7.1 gives more detail of electrical motor and generator capabilities.

The striking information from Figure 7.1 is how low the power/weight ratios of electrical machines are and how much better for the renewables industries if those power/weight ratios were higher.

However, just as there have been large improvements to heat engines, in the transition from reciprocating to turbine drives.

That revolution also needs to come for the electric machine, and it is occurring in the development of electric drives for vehicles and particularly aircraft, whose applications demand much higher power-to-weight ratios.

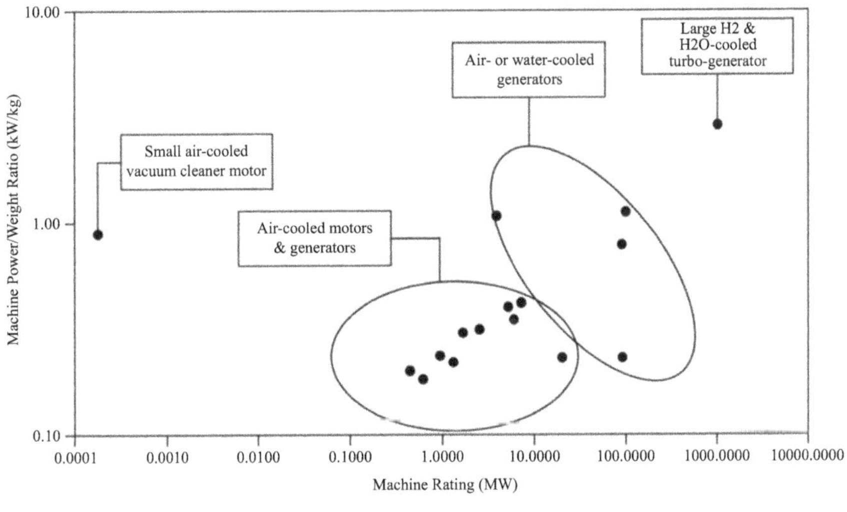

Figure 7.1 Graphic of electrical machine power–weight ratios against their rating

Table 7.1 Examples of rotating electrical machines, typical sizes and loadings

Type of machine	**DC permanent magnet motor**	**Air-cooled 3-phase AC induction motor**	**Air-cooled 9-phase advanced AC induction motor**	**Air-cooled 3-phase AC synchronous hydro-generator**	**Air-cooled 3-phase AC synchronous turbo-generator**	**H_2O and H_2-cooled 3-phase AC synchronous turbo-generator**
Cooling type	TEFC	CACW	CACW	CACW	CACW	CHCW
P, MW	0.0002	6.14	24	90	100	1,000
Poles	4	4	12	88	2	2
Rev/min	6,000	1,790	120	68.2	3,000	3,000
Typical weight, tonnes	0.0002	18	89	400	90	350
Power/weight, kW/kg	0.88	0.34	0.22	0.23	1.11	2.86
Magnetic loading, air-gap flux density, T_{rms}	–	0.90	–	1.05	0.94	1.00
Electric loading, A/mm	–	93.0	–	62.5	116.3	255.3
Mechanical loading, rotor peripheral velocity, m/s	–	69.5	14.1	46.3	139.8	188.5
Air-gap torque density, kNm/m^2	–	140.6	707.4	140.6	1,518.4	736.8
Current density, A/mm^2	–	3.9	3.9	3.7	2.8	11.1

This revolution will be based upon the use of better materials and better designs to achieve much greater power-to-weight ratios than those shown in Figure 7.1 and Table 7.1 but also to achieve lower weights and manufacturing costs.

Current research in the United Kingdom by the Future Electrical Machines Manufacturing Hub is attempting to increase power/weights significantly and Jackson and Mecrow [32] have produced an *Electrical Machines Road Map*, with the following targets from 2020:

- Current power to weight ratios, 0.2–3 kW/kg;
- By 2015 to achieve a power to weight ratio, 7 kW/kg;
- By 2023 current information suggests some high-performance motors for electric drives are progressing from 3 towards 7 kW/kg.

7.3 Early AC and DC networks

7.3.1 Transformers and AC transmission development

The earliest AC transformers were being constructed towards the end of the nineteenth century in Europe by Ferranti and in the USA by Westinghouse and Tesla. Other designers in Europe were also developing transformers because of their innovative work to develop the long-distance transmission of electrical energy.

This followed Edison's development in the USA of DC transmission and its concentration in urban areas because a lack of technology for long-distance transmission.

The application of AC transmission quickly expanded the development of long-distance AC networks, particularly in the USA, because of the transmission distances required, but also in Europe, see Figure 7.2, where it stimulated the industrial expansion of France, Germany and the United Kingdom, following the lessons of the 18th and 19th centuries and their explosion of scientific information.

7.3.2 Spread of technology, Europe and USA to rest of the world

These scientific lessons quickly galvanised the World so that from 1900 electrical experiments, the building of power stations and delivery of power to houses and factories started to grow all over the developed World, in China, Japan, India, North and South America, see Figure 7.3.

7.4 Switching and control

7.4.1 Switches

An early issue for controlling AC or DC networks was the ability to switch power on and off. The earliest electrical switches were mechanical and required rapid

(a)

(b)

Figure 7.2 *Examples in Europe of AC transformer and electrical power transmission development. (a) Early Oerlikon 3 phase, 40 Hz, AC transformer in 175 km power transmission line from Lauffen am Neckar to Frankfurt, Germany, 1891 (b) First European 110 kV transmission line in Germany, between Lauchhammer and Riesa, 1912.*

operation to ensure that the extinction of the consequent electrical arc, particularly severe for the interruption of DC circuits.

7.4.2 Interrupters

An interrupter can be a knife blade switch to interrupt electric current abruptly.

The device will have both a manual operating mechanism, combined with an automatic operation, triggered either by the magnitude of current flow through the interrupter or intervention from a remote operator.

The problem for an interrupter is that, as load current increases, it becomes more difficult to complete current interruption without thermal or electrical damage to the blades or switch mechanism.

The solution is an interrupter with a very powerful mechanical spring or electromagnetic mechanism to accelerate the switch blade separation. The blades must be constructed of materials that can withstand high temperatures and the radiant thermal power of any electrical arc produced.

A DC interrupter has a more difficult job to clear a DC current which remains constant as the interrupter is opening. This can be achieved by strong and very

Figure 7.3 Examples of the rapid spread of electrical technology from Europe to N and S America, China, Japan and India. (a) Brush Electric Company's Central Power Plant, DC dynamos power New York public arc lamps, 1880; (b) Nanjing Road, Shanghai, China, electric street light illumination, 1883; (c) Keage Power Plant, Japan, Lake Biwa Canal first hydro-electric power plant, 1890–97; (d) Bangalore, first Indian city to install electric streetlights, 1905.

fast-acting operating springs combined with a spacious design that extinguishes the current arc.

An AC interrupter has an easier job as the AC current varies sinusoidally and drops to zero twice every cycle, so an AC interrupter can afford to be more lightly constructed and be more compact than a DC interrupter.

AC has the great advantage of permitting the transmission of large quantities of electrical power over long distances, without difficulties of interruption as Dc experiences.

7.4.3 Circuit breakers

A hierarchy of circuit breakers is shown in Figure 7.4, AC devices to the left, the few DC devices on the right.

To enable the safe operation of high-power AC electrical circuits powerful electro-mechanically operated circuit breakers were developed and a wide variety air-break, oil-break and vacuum, break breakers have evolved for specific technical requirements.

A typical example of three phase AC air-break Circuit Breakers are shown in Figure 7.5, installed in a Finnish Power Station output switchyard.

The powered circuit breaker is an essential electromechanical technology for safely interrupting AC circuits from 30 up to 1,500 kV, handling the breaking of currents up to 1,000 Amps, but that is relatively straightforward in an AC circuit where the current falls to zero twice every AC cycle.

DC circuit breakers have greater difficulty interrupting DC current as the current remains constant and does not pass through zero, as an AC current does. This interruption difficulty becomes especially problematic as DC circuit current loads increase.

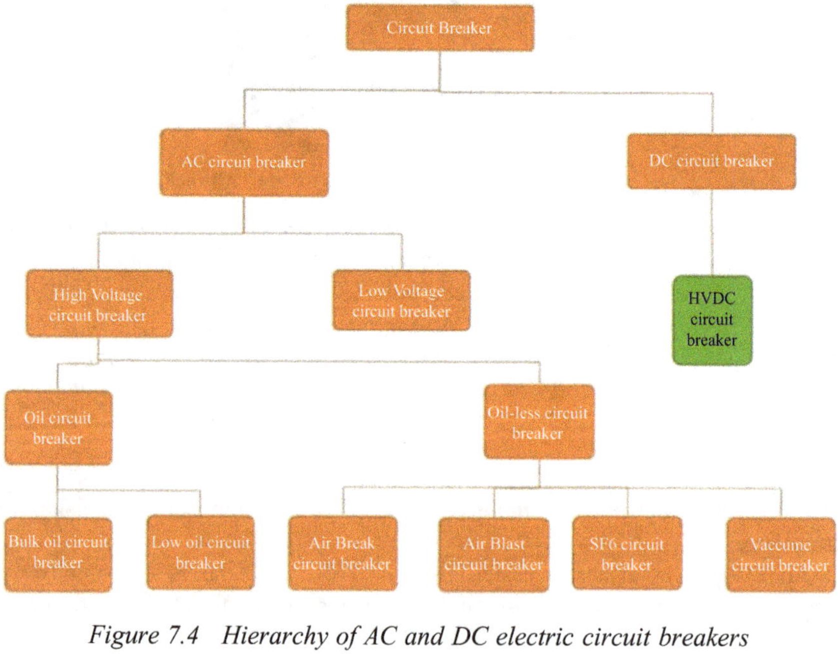

Figure 7.4 Hierarchy of AC and DC electric circuit breakers

Figure 7.5 Substation feeding the AC Grid from a Finnish Power Station, with 3 phase HV AC circuit breakers capable of interrupting fault currents

However, DC power has some advantages over AC, in that it can be more easily controlled by semiconductor switching, without the additional costs of intervening transformers and associated circuit breakers.

DC power could therefore be more amenable to the expansion of renewable sources of energy, and DC transmission techniques could be valuable for the future, but to handle DC power appropriate valve and semiconductor switches, to interrupt DC currents, will be essential.

The switching of electrical power advanced in the twentieth century with the development of electronic valves and their modern replacement, semiconductor switches, Figure 7.6.

7.4.4 Valves and semiconductors AC/DC/AC conversion

The simplest switch is the diode, which can rectify a sine wave, Figure 7.7(a), the thyristor enables the rectification and control of a sine wave, while the IGBT (Inverse Gate Bipolar Transistor) enables the forming of an AC sine wave from a DC voltage, Figure 7.7(b). See Figure 7.7, the simplest switch is the diode, which can rectify a sine wave, Figure 7.7(a). The thyristor enables the rectification and control of a sine wave, while the Inverse Gate Bipolar Transistor (IGBT) enables the formation of an AC sine wave from a DC voltage, Figure 7.7(b).

Device	Symbol	Semiconductor structure and features	
Diode	Anode  Cathode	One PN junction: rectifier function (automatic switch) 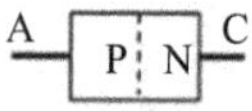	
Thyristor	Anode Gate Cathode 	Three PN junctions enable controlled-rectifier function 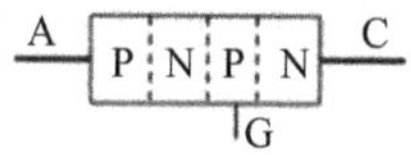	
Bipolar transistor	Collector Base NPN Emitter 		Low input impedance is a drawback, but the collector-emitter current capacity is large when the base is used as the input terminal with respect to the emitter.
	PNP 	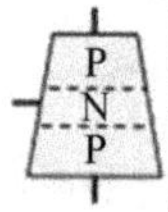	Base (B) Collector (C) Emitter (E)
MOSFET	Drain Gate Source 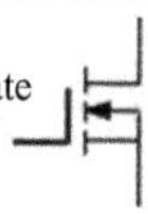	Made possible by semiconductor microfabrication technology. Impedance of gate input is high. Works well with logic circuits.	
IGBT	Collector Gate Emitter 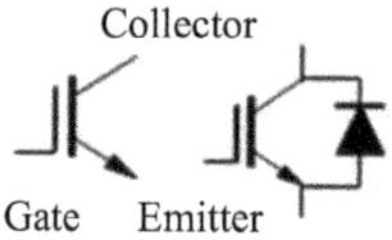	Combines high input impedance of MOSFET and large current capacity of bipolar junction transistor. The device on right incorporates a reverse-parallel diode.	

Figure 7.6 Examples of various semiconductor switches, capable of chopping DC current, enabling conversion of current from DC to AC and AC to DC

7.5 HV AC and DC advanced networks

7.5.1 Electrical EHV and HV transmission

Figure 7.8 shows the current structure of Great Britain's National AC EHV and HV Grid, including interconnections both to Ireland in the West and other European countries to the East, via HVDC cables.

7.5.2 AC or DC networks

Modern AC electrical networks now extend of over large areas of the Earth, as will be shown in the next Chapter 8, particularly refer to Figures 8.13–8.15.

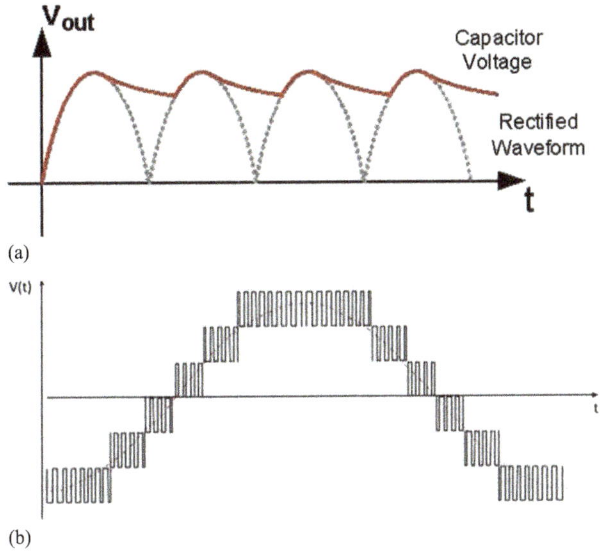

Figure 7.7 Examples of conversion AC–DC or DC–AC. (a) AC–DC diode bridge rectification, after capacitor smoothing; (b) DC–AC IGBT bridge rectification, prior to capacitor smoothing.

However, modern DC networks are gradually becoming more common, as reliable performance high voltage Thyristor and IGBT Converters were developed to interconnect isolated islands and maritime countries developed and new renewable sources are being connected by DC links to widespread AC electrical networks.

An example of this DC interconnectivity in Europe is shown in Figure 7.9.

Most HVDC links transfer power between isolated national networks or from renewable sources such as hydro and wind.

7.5.3 Cables and losses

As electricity is transmitted through transmission lines or cables, see Figure 7.10, there are consequent I^2R losses which abstract energy from the line.

Suspended transmission lines, Figure 7.10(a), are cooled by their suspension in air but buried cables, Figure 7.10(b), cannot discharge that heat loss, except radially through the cable wall or by coolant circulated through pipes in the cable core, which increases the cable cost.

However, that coolant technique is commonly used for EHV cable systems, particularly around large cities, for example the EHV cable ring around London is water-cooled.

7.6 Interconnectors, storage and big grid stability

Alternating current electrical system networks must be stabilised that means that the energy injected into the network must match the energy demanded, minute by minute, second by second.

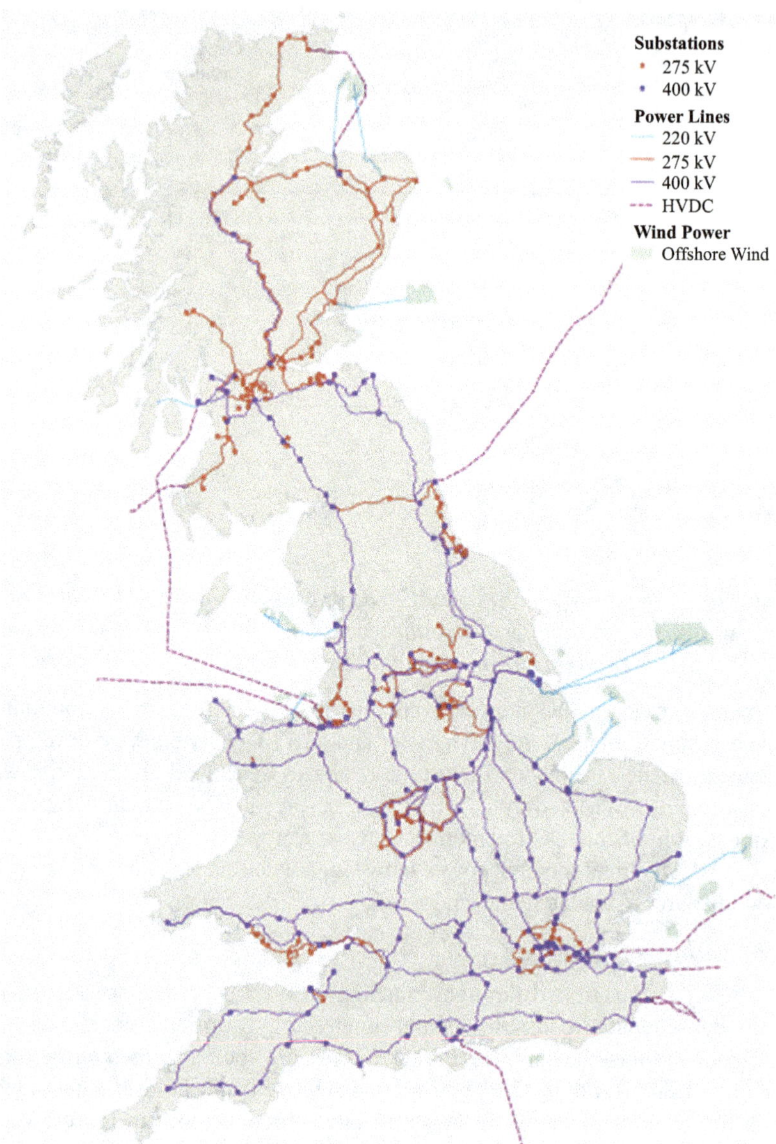

Figure 7.8 Great Britain's 400 and 275 kV AC National Grid, with AC links to some offshore wind farms, DC links to some distant offshore wind farms and HVDC links to Ireland, France, Belgium, the Netherlands and Norway

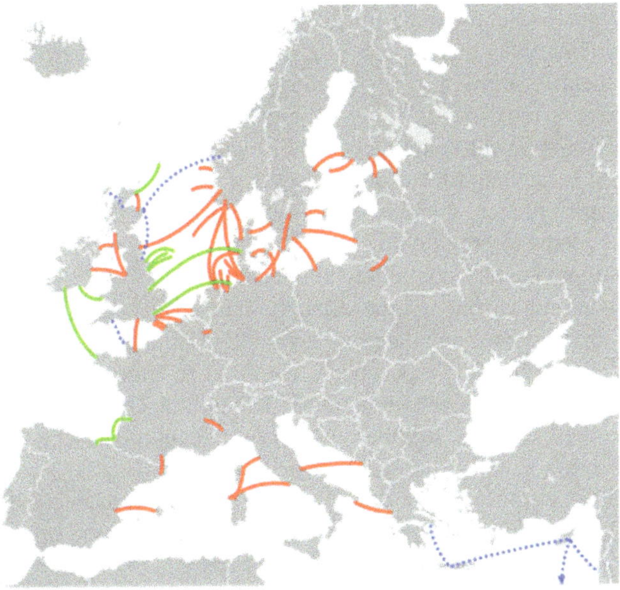

Figure 7.9 *European DC Electrical Interconnections. Expansion of Links in Baltic, North, Irish and Mediterranean Seas; (a) red: existing links; (b) green: links under construction; (c) blue: proposed links.*

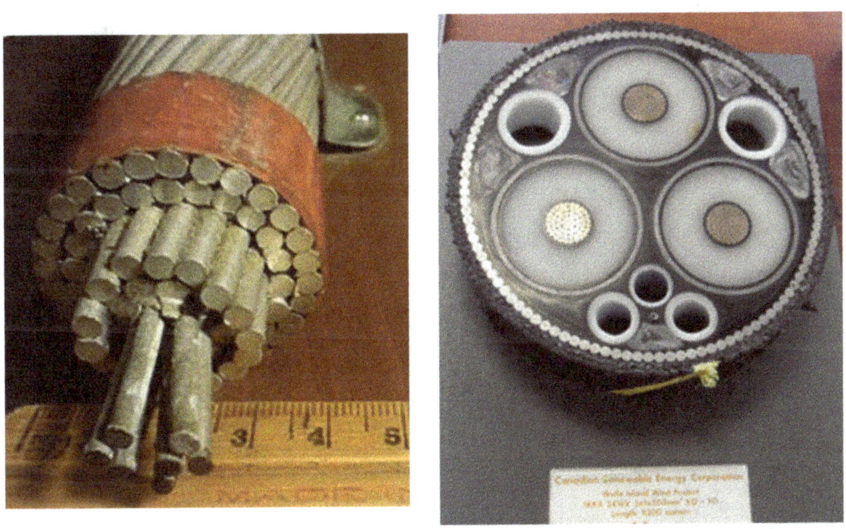

(a) (b)

Figure 7.10 *Examples of AC cables, suspended and buried. (a) Section through a suspended Aluminium Conductor Steel-Reinforced (ACSR) EHV transmission line, note the steel core strands; (b) section through a sheathed, buried 245 kV three-phase submarine power cable, note the cooling ducts.*

So as more or less energy is demanded the electrical network must be balanced. The basic way that is done in a single AC network is by governor speed control of generator prime movers, rotating at the demanded frequency of 50 or 60 Hz. Governors must adjust steam, gas turbine or diesel driven prime mover power to match the power being demanded by the grid.

Power can be supplied from storage systems within the grid. Some European countries, notably France, Germany, Norway and Grand Britain have developed hydro-electric plants to provide that stability.

Late twentieth century control technologies have allowed some hydro-electric plants to provide adjustable storage capacity akin to that of very large batteries. Examples of this will be shown later in Section 9.2.5, and these can help to balance and stabilise electrical grids, especially when they are larges.

It is interesting that in a large country, China, the Zhōngguó Gòngchǎndǎng government (中国共产党), originally planned a country-wide network of AC transmission lines to cement the whole nation.

However, practicality meant that this eventually evolved into the aggregation of three separate but unsynchronised 50 Hz AC systems, based upon three urban centres, Beijing, Hong Kong/Guangzhou and Shanghai.

These networks have been interconnected by short back-to-back AC/DC/AC interconnectors and has included the deployment of high-power AC/DC/AC converters and the deployment of large synchronous condensers or synchronous compensator, within the AC networks, to adjust the power factor within the transmission grid (Figure 7.11).

The synchronous condenser is a DC-excited synchronous generator, whose shaft is un-connected to a prime mover but spins freely; therefore, it does not generate or absorb MW.

Figure 7.11 Synchronous condenser used to stabilise an AC Grid network

Its purpose is not to convert electric power to mechanical power or vice versa, but its field is controlled by a voltage regulator to allow the synchronous condenser to generate or absorb reactive power, MVArs, as needed to adjust the grid's voltage, or to improve power factor.

The practical and rapid development of these interconnectors has meant that China's 3 independent AC systems are now interconnected, supported by synchronous condensers and effective pumped storage from large river hydro-electric schemes, thus ensuring that China's very large AC electrical system has greater resilience.

7.7 Distributed low energy electricity

7.7.1 MV electricity distribution

Medium voltage electrical networks in cities, towns and nearby countryside are the backbone of electricity transmission, providing power and lighting for homes, schools, small businesses and larger enterprises.

These networks enable an economy to thrive but in developing countries, in Africa, South America and South East Asia MV networks are difficult to maintain and sustain. This is partly because of the cost of building the MV network infrastructure, but more because of the demands in maintaining its operational capacity to deliver electricity, due to many users wanting to access the network who cannot afford to pay for it, so in effect illicit users piggy-back on the system, causing network voltage to collapse and all users to lose the service. This is a common experience in developing countries with nascent electrical systems.

So, building MV networks requires discipline and control but also a common goal that there is more value in supplying reliable customers than in trying to stretch the service beyond its electrical limits.

An author's 1973 vignette on this issue was from personal experience working in Benin City, Nigeria, where the National Electric Power Authority (NEPA) provided a strong, 4 spoke 11 kV system for the city. However, illegal actors threatened to destroy its benefits by abstracting power from the lines, with crocodile clips and wires, without paying for the service, a situation that is difficult for local network engineers to manage.

7.7.2 Photovoltaics, wind and hydro

It is likely that photovoltaic panels on buildings can provide a valuable energy source for local, grid-connected MV networks but that requires a control system and perhaps a storage battery to match the local network voltage to any PV cell output transformer voltage and a degree of automatic control will be required to achieve that. Installations have grown in locations with high insolation, but the investment is costly and returns modest, so this has not yet been a highly successful market, but perhaps in countries with less reliable grids, for example in Africa, South India or South-East Asia, maybe it will find an effective niche.

Figure 7.12 Mixed source renewable AC Grid in Scotland on the Isle of Eigg, with a view of the PV panels

There has also been great enthusiasm for small wind power turbines, generating less than 1 kW, to be mounted on houses. Again, the coupling to a local AC network is important, requiring storage and control.

Certainly, this option is expanding, and many devices have been installed in windy offshore islands with low power electrical grids.

In Scotland on the Isle of Eigg, a successful AC electrical mini-grid has been developed. Eigg is not connected to the mainland electricity supply, Figure 7.12. After decades of relying on diesel generators in 2008 Eigg Electric developed a 24 h, community owned, managed and maintained electrical power system, providing electricity for all island residents from renewables incorporating the following sources:

- Hydroelectricity, one large loch turbine capable of generating up to 100 kW, with two smaller stream turbines 5–6 kW turbines;
- Wind power, four small 6 kW wind turbines;
- Photovoltaics, an array producing 170 kW from the sun.

7.7.3 Distributed network tools

7.7.3.1 Introduction

A distribution network tool is software that can be used to analyse and improve the operation of 3-phase distribution networks or Smart Grids.

The range of tools available begins with the determination of the optimal tie open point for a distribution network to minimise loss or reliability indices, whilst still complying with imposed network constraints.

The software tool continues to optimise a distribution network voltage profile during operation, with the objective of preparing for increased numbers of distributed energy sources or an increasing numbers of loads within an LV Distribution Network through the estimation of optimal distribution transformer tap positions within the network.

For unbalanced network conditions, a 3-phase balance optimisation algorithm can used to find the optimal balance, for a given load and available generation units, between the three phases.

Some software packages may also include tools for optimising the placement of capacitors within a network to maintain acceptable overall power factors.

Hosting capacity tools can also be used for evaluating the maximum distributed energy resources needed by a distributed network and to estimate the spare load capacity of a network.

7.7.3.2 Telephone, telecoms and computer controls

Electrical distribution networks are becoming increasingly complex systems.

The addition of embedded generation, sustainable energy projects, demand-side management schemes and participation smart metering require system operators to gather unprecedented levels of information from their network in real-time.

Current SCADA technologies, such as power-line carrier or slow scan UHF telemetry, have a low throughput or bandwidth and are overwhelmed, given the task of providing service data for the volume of devices needing telecoms predicted, given the volumes to be installed on MV and LV power systems in the medium term.

Alternative and emerging solutions are being investigated to provide telecoms access to remote devices, with a focus on IEEE 802.16 WiMax. WiMax has been gaining momentum over the last few years as a solution in fixed 'last mile' environments, providing broadband internet access to rural homes and businesses, as it provides excellent geographic coverage for a reasonable capital investment.

Telecommunications are now an essential component for emerging Smart Grid technologies.

Up to date, power system telecommunications planning has previously focused on EHV and HV transmission with less investment in MV and LV distribution networks.

There is a need now, with the growth of renewable energy power sources feeding into MV and LV distribution systems, for a universal system over which all messaging may be achieved.

With the rise of protocols such as IEC 61850, an international standard defining communication protocols for intelligent electronic devices at electrical substations, there is growing momentum to install telecoms to exploit the control benefits on offer. However, the challenge is to correctly engineer systems that can accommodate rapid growth, whilst catering for throughput demand and real-time performance.

Some of these deal with last-mile access technologies or backhaul solutions, proposing a model telecoms network for Smart Grids.

7.7.3.3 Metering and billing

This concerns the incentive problem in MV and LV electricity distribution when customer energy usage cannot be perfectly observed by the utility supplying the energy.

It is assumed that each customer has private information about the amount of their consumed energy but this information is not available to the supplying utility. Imperfect observability of individual user demand results is non-technical energy losses.

In developing countries, these losses can amount to 20%–30% and are largely attributed to theft by residential customers.

Reducing such losses would allow a marked increase in electricity distribution efficiency.

The use of smart energy management devices would enable new Smart Grid functionalities with the potential to increase efficiency in the use of renewable energy sources. However, employing smart energy management devices also entails problems.

Typically, such devices are usually commercially produced, employ off-the-shelf information technology solutions, with inherent security vulnerabilities.

However, due to cost constraints and their technology limitations, such smart devices are tamper-vulnerable and may enable systemic energy theft, threatening to reduce or even erase any efficiency gains.

To address these issues, there need to be incentives for the utility supply company to combat theft, while the utility may be subject to National or Regional tariff regulations.

Experience shows that such regulated utility supply companies invest less than is socially optimal for theft reduction. It is suggested that regulators should include explicit targets for the allowable losses, to remedy the problem of incentive misalignment in smart device energy management.

7.8 Summary

This chapter has concentrated on the significance of electricity, electrical machinery and power transmission to facilitate cleaner energy and its increase.

Initially this was achieved through electrification but more recently by facilitating the injection of renewable energy technologies into the electrical grid at lower voltages.

The chapter has described the late nineteenth century development of DC power generation in the USA by Edison, closely followed, in the same decade, by AC power generation in both United Kingdom and USA pioneered by Ferranti, Westinghouse and Tesla. The latter development was followed by the rapid expansion of long-distance AC transmission due to AC technology's superiority, in terms of range and capacity, over earlier DC networks. In the next few years, into the twentieth century, AC networks proliferated, initially in the USA and Europe, then more extensively in the USSR, Russia and more recently in China, East Asia and South America.

AC networks, with their 3-phase transmission lines and substations, have developed strongly over the last 120 years, starting at MV voltages from 11 to 33 kV but steadily increasing in voltage over time to HV and EHV voltages up to 1,500 kV, with line lengths ranging from a few hundred km to thousands of km. These systems grew in the twentieth century, initially expanding in industrial cities but also because of growing domestic electricity usage, followed by world-wide industrialisation reaching remote locations over longer transmission distances.

Exemplars for this have been the electrical interconnection of societies and countries, initially in North America, Europe, Russia, India and then South America but since Second World War in China and East Asia and most recently in the countries of Southeast Asia and Africa.

In the twenty-first century these MV, HV and EHV AC transmission networks have also shown themselves capable of absorbing variable renewable power sources, so that AC transmission networks are forming an important role as a back-bone for future renewable energy supply.

The development before Second World War of high-speed vacuum valves, stimulated by the development of pulsed radar, was rapidly followed by the invention of even faster semiconductor switches, enabling the development and growth of HVDC transmission networks. These networks have facilitated lower cost, lower loss, longer distance DC electrical power transmission links, first for electrical supply between important islands in the Baltic Sea and the interconnection of onshore AC electrical systems of different frequencies, such as in Japan, but then for the widespread interconnection of AC networks of nominally the same frequencies, as in China.

This chapter has explained the development of the following technologies, valuable for the expansion of renewable energy by electrical means:

- MV early power networks.
  - The expansion and interconnection of city and regional MV AC networks in the early twentieth century;
  - Their expansion and interconnection as national MV AC networks in the mid-twentieth century;
  - The importance of improved control and security of MV grid systems;
  - Recent influence of these controls on the ability of MV grids to absorb and transmit variable renewable power.

- HV and EHV national and international electrical transmission systems:
  - Expansion and interconnection of EHV and HV energy power networks and HVAC grids;
  - Growth of HVDC interconnectors between AC grids;
  - Both enabled by HV and EHV power electronic converters in those interconnectors;
  - Continental and inter-continental expansion of electrical power systems using HVAC grids connected by HVDC interconnectors;
  - The influence of the injection of variable renewable power sources into those extensive networks;
  - Control of these larger interconnected EHV systems and their ability to accommodate large swings in injected variable renewable resources.

The chapter completes by describing the distribution of lower electrical energy through MV and LV circuits, describing the contribution to this of the interaction between power and communication networks.

The consequential importance of storage and computation for maintaining the stability of large interconnected complex EHV, MV and LV electrical networks with variable renewable energy.

Chapter 8

Modern age IV, 1970–present CE, Energy Vectors, storage and networks

8.1 Introduction

This chapter returns to Sadi Carnot's thermodynamic ideas, describing the concept of heat as an Energy Vector.

In the eighteenth century, this idea improved our consequential understanding of the energy obtainable from fossil fuels, coal, oil and gas, as Energy Vectors, an understanding of their ability to raise temperature, deliver heat and power, enable transportation and the efficiencies of these thermodynamic processes.

The thermodynamic ideas were followed in the nineteenth century by the development of electrical thought, through an understanding of electromagnetism, bringing the benefit of the Electrical Energy Vector, with its potential for universality and control, making possible the distribution of power to towns and cities and, through long-distance transmission lines, across countries and continents.

These ideas altered former ideas of fire and motion, see Figure 8.1, encouraging us scientifically forward into a thermodynamic and electric age.

This chapter will try to demonstrate some of the benefits of these Energy Vectors and networks, whether they are electrical power lines, oil or gas pipelines, roads, railways, ships voyages, telephony, the internet or satellite links, demonstrating the extent, size and impact of some of these innovations.

The chapter will also introduce the mid-twentieth century technology of nuclear power, its early rapid expansion, consequent learning curve and accidents, and the impact it has had, particularly in countries without the benefits of fossil fuel resources.

As part of the introduction to this chapter, refer to Chapter 3, Figure 3.1 and the impact of the Sun's energy output on Earth.

Based on this, Figure 8.2 shows the Sun's energy input to Earth and the 2012 current World electricity demand.

(a) (b) (c) (d)

Figure 8.1 Iconographies of fire and oil, rationality and electricity. (a) Gustave Moreau's Prometheus, bearer of fire. (b) Trans-Alaska Oil Pipeline 1,287 km long, 1.22 m diameter. (c) William Blake's Isaac Newton, bearer of rationality. (d) US EHV 3-phase 500 kV Transmission Lines Grand Coulee Dam.

It is clear from this graphic that this demand could be met by extracting energy from the Sun's incoming power, using solar PV, wind, wave or tidal energy, if we were able to deploy those technologies to sufficient size to extract that energy.

This demonstrates that human development does not need to rely solely upon fossil fuels to meet energy needs.

Note that this analysis excluded the extraction of energy from any nuclear sources.

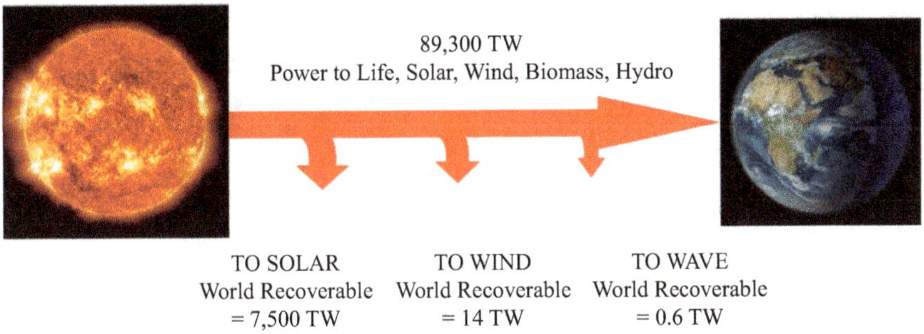

THE SUN	89,300 TW
Solar Power, Potential	7,500 TW
Wind Power, Potential	14 TW
Wave, Potential	0.6 TW
Tidal, Potential	0.2 TW
Current World Electrical Demand (2013)	2.2 TW

TO TIDAL,
not based on Solar flux
but to gravitational field
between Moon & Sun & Earth,
Recoverable
= 0.2 TW

Sources: Tim Grinsted & SANDIA Report

Figure 8.2 Earth's insolation and the Sun's energy contribution

8.2 Energy vectors and networks, significance

8.2.1 Heat as an Energy Vector

Savery, Cugnot and Newcomen the builders of the early steam engines of the eighteenth century understood the action of steam on the piston and the value of condensing that steam to return the piston to its starting position.

But they had relatively little mechanics knowledge and no understanding of thermodynamic processes or the benefits that such knowledge could confer.

Their steam engine designs had thermal efficiencies of less than 1%.

It was Watt who combined an understanding of thermodynamics, from Clausius and Clapeyron, with his practical machinery skills of levers, motion, optimisation and thermodynamics, learnt as a watchmaker, machinist and steam engineer.

This knowledge enabled him to develop far superior steam engine efficiencies and for us to understand the steam engine's performance in terms of a Heat Energy Vector.

It is these combinations of science and technology, which has enabled modern efficient steam reciprocating and turbine engines possible, ultimately achieving thermal efficiencies expanding from 14% to 40%, overtaking the performance of any of the earlier steam engines.

It was the thorough steam engine mechanics and thermodynamics that then kick-started the modern developments of petrol, diesel and subsequently gas engines.

8.2.2 Thermal storage and networks

For large steam applications, in ship propulsion and electricity generation, huge steam-raising boilers, extensive pipework systems and huge condensers became essential to achieve efficiencies.

This required an understanding of high-quality insulation and well-designed thermal networks to ensure the highest efficiencies.

Some understanding of thermal storage was also needed to limit the wastage of heat from large and extensive industrial processes, such as steel- and paper-making, making use of thermal stores to improve the overall plant thermal efficiencies.

8.2.3 Fossil fuels as Energy Vectors and their calorific values

Steam cannot be raised, nor ignition commence without the availability of a fuel to be burnt to release energy.

The World's most readily available sources of that energy are the fossil fuels, coming from the Earth's surface and crust, with varying calorific values summarised in the table below.

Methane gas	55.5 MJ/kg ≡ 15.4 kWh/kg
Oil, depending on type	47.0 MJ/kg ≡ 13.4 kWh/kg
Diesel fuel	44.8 MJ/kg ≡ 12.4 kWh/kg
Anthracite coal	32.5 MJ/kg ≡ 9.0 kWh/kg
Sub-bituminous coal	23.9 MJ/kg ≡ 6.7 kWh/kg
Moisture and ash-free firewood	21.7 MJ/kg ≡ 6.0 kWh/kg
Lignite coal	17.4 MJ/kg ≡ 4.8 kWh/kg
Dry peat	15.0 MJ/kg ≡ 4.2 kWh/kg

These figures show the relative benefits of fossil fuels, particularly of gas, clearly the most energetic and of dry peat, the least energetic.

However, from the eighteenth century, the consequences to the Earth's atmosphere of the widespread burning of fossil fuels had gradually become more apparent.

8.2.4 Electricity as an Energy Vector

The invariable early sources of energy to generate the Electrical Energy Vector were the fossil fuels described above.

They were burnt initially to raise power for reciprocating- or turbine-driven rotating machinery, to generate electricity, initially in the nineteenth century using coal to raise steam, but later in the twentieth century using oil or gas for steam-raising. The coal-fired transatlantic liner RMS TITANIC burning coal to power 2 reciprocating engines and an inline steam turbine had, for its day, one of the highest thermal efficiencies of any heat engine and certainly pioneered the thermodynamic efficiencies for the forthcoming ship, electrical and aviation ages.

So, Thermodynamic Energy Vectors, coal, oil or gas, burnt to generate heat and electricity, were progressively transforming, by heat engines and electric generators, into the Electrical Energy Vector.

This transformation to either DC or AC power supplies has important implications, in terms of control, described briefly in the following sub-sections.

8.2.4.1 DC voltage and current control

Using Edison's technology, the earliest electrical generators were DC.

These machines had DC field magnets mounted on the dynamo's stator, surrounding a rotating wire-wound rotor armature and commutator.

The rotor contained windings in which an AC voltage is generated, due to the rotor's rotation in the air-gap's DC magnetic field.

This armature AC voltage was then rectified to DC by the rotor commutator and brushes, the DC output being a DC Electrical Energy Vector.

Control of that Electrical Energy Vector was achieved by adjusting the current supplied to the generator field magnet windings.

Initially, in the nineteenth century, this control was exercised manually, by an operator controlling a resistor, adjusting the field current dependent on the generator's output voltage.

Relatively soon rapid automatic rheostatic control was developed to maintain the DC dynamo output voltage constant, but adjustable remotely from a control room.

8.2.4.2 AC voltage and current control

AC electrical generators, on the other hand, rapidly superseded the DC transmission technology, using Ferranti's methods further developed by Westinghouse and Tesla.

These AC generators had DC field magnets mounted on the rotor, energised alternately positive and negative, North and South, surrounded by a stator containing a stationary series of AC wire-wound phases to complete the machine's structure.

The rotor induces AC voltage into the stator windings, due to the rotation of the DC field magnets with respect to the stationary AC armature.

AC currents of varying phase were then drawn from successive phases of the AC stator armature to become the 2-, 3- or even 4-phase AC Electrical Energy Vectors.

Control of output was achieved by adjusting the voltage supplied to the generator rotor field magnet windings, dependent upon the AC voltage appearing at the stator winding terminals.

Again, initially this control was initially exercised manually, by field current adjustment, based upon the required generator output voltage. This was again soon superseded by an automatic voltage regulator (AVR) to maintain generator AC output voltage constant, also adjustable remotely from a control room.

These AC systems of Ferranti, Westinghouse and Tesla then rapidly overtook Edison's DC system, because of the simple fact that, by the use transformers,

AC voltages could be adjusted over a wide range to meet differing electrical power demands and transmit power over huge distances. For example,

- From low voltage (LV) for domestic consumers;
- To high voltage (HV) for industrial users;
- To extra HV (EHV) for long-distance power transmission.

These transmission lines, raised to very HVs, carried very low currents, enabling power transmission over very long distances with very low losses.

It is interesting to reflect now that Edison (1847–1931), one of our greatest scientists and inventors, devising many ingenious inventions and laying the foundations of modern electrical science, was in terms of electrical engineering completely overshadowed by Ferranti (1864–1930), Westinghouse (1846–1914) and Tesla (1856–1943), because of their advocacy of AC over DC electrical transmission.

8.2.5 Energy Vectors, actual and virtual storage mechanisms

Most energy production and transmission methods contain some element of storage, ranging from capacitors through Li/Fe batteries to huge upland reservoirs, although some storage capacities are small, they are not necessarily insubstantial or useless to benefit the electricity system.

However, others have proved extremely valuable for maintaining energy supplies as energy demands change, sometimes rapidly.

The following indicates how different electrical energy supplies can meet varying demands in the electrical power grid or in trucks, cars, trains, ships and aircraft. For example,

- Hydro-electric generation:
  Implicitly based on large water storage reservoirs, so having large and significant storage capability.
  However, hydro-electricity is dependent upon river flow-rates, which in turn depend on the seasons and weather conditions, which may limit their storage capacity during some parts of the year.
- Fossil fuel fired power stations:
  These can deal with relatively rapid changes in demand when they are operated in large fleets, being achieved by larger more efficient stations operating in battery formation to meet the fixed base load demand, whilst other, smaller more agile, power stations are able to make relatively rapid variations to respond to changes in demand.
- Nuclear power stations:
  These use long-term slow reactor controls to deliver the electrical power demanded by the electrical transmission grid.
  The safety implications of tight reactor control mean that nuclear plants cannot supply rapid power output changes to the grid. Rapid power changes

are limited to protect the reactor and its nuclear fuel from short-term nuclear process contaminants, which can affect long-term reactor performance.

- Diesel engines:
  These prime movers can deal with rapid demand changes, demonstrated by their effectiveness as heavy truck and railway locomotive motive power units. But there have also been many improvements in diesel engine fuel injection and designed weight to achieve this agility.
- Petrol engines:
  These engines can also deal with rapid demand changes. In some respects, they are more responsive to a sudden power demand change than a diesel engine. That is why they have proved so popular for motor car propulsion, particularly those of high performance.
  There have been many design improvements in petrol engines, similar to diesel engines, particularly in engine weight, fuel injection and controls to achieve these effects.
- Gas turbine engines in aircraft:
  The gas turbine first appeared in aircraft because of its low weight and high power output. Gas turbines high power-to-weight ratio is achieved at the cost of using a high light paraffin fuel. The aero gas turbine can respond rapidly to changing power demands but cannot provide storage.
  However, most large passenger aircrafts carry a small gas turbine or auxiliary power unit to provide rapidly variable electrical power, required particularly during take-off, in turbulent conditions and on landing.
- Gas turbine engines in ships:
  The designs of ship gas turbines differ to those powering aircraft. They are more rugged units with power-to-weight ratios much lower than for an aero engine. That means that they are capable of releasing some energy which can be stored, from their heavier diesel fuel, for example into a ship's electrical power system, hot water or air conditioning systems.

8.3 Nuclear power

8.3.1 Introduction

Work on the fundamental physics of matter had developed in the United Kingdom, particularly at Cambridge and Manchester Universities in the late nineteenth century and early twentieth century, some of it was inspired by Marie Skłodowska-Curie's work and was led in the United Kingdom by two later outstanding physicists, Figure 8.3:

- Joseph John Thomson, credited with the discovery of the electron, the first subatomic particle to be identified;
- Ernest Rutherford from New Zealand, who was a pioneering researcher in atomic and nuclear physics, described as 'the father of nuclear physics' and 'one of the greatest scientists of all time'.

(a) (b)

Figure 8.3 Founding discoverers of the atom. (a) Professor J J Thomson, 1856–1940, Cambridge University (b) Professor E Rutherford, 1871–1932, Canterbury, McGill, Cambridge and Manchester Universities.

8.3.2 Early nuclear fission power

Nuclear power developed out of the experiences of the Second World War and the development of nuclear weapons, first used by the USA against Japan in 1945, which brought the war to an end. Many countries had an embryonic nuclear weapons programme, including Germany, UK, USA, Soviet Union and Canada. On the Second World Wars conclusion, some of these countries started programmes to develop peaceful outputs from their nuclear weapons programmes and a variety of nuclear power technologies developed.

A nuclear reactor can initiate and control either a fission chain or fusion reaction. The fission reactors described earlier use heavy elements, such as uranium or plutonium, whose atoms are split by neutron bombardment releasing fission energy. The fission process is moderated by neutron absorbers such as a graphite carbon core or by water in a containment vessel. Heat from nuclear fission is transferred to the working fluid, gas or water, which is used to raise steam, powering steam turbines to drive electrical generators.

The steam generated from a nuclear reactor could, in principle, be used for industrial process heat or for district heating, but that application has not yet proven viable.

Some nuclear fission reactors are used to produce isotopes for medical and industrial use, or for the production of weapons-grade plutonium.

In 2022, the International Atomic Energy Agency reported that there are 422 nuclear fission power reactors and 223 nuclear research reactors in operation around the World.

The United Kingdom was the World's first country to install nuclear power stations for the generation of electricity, and in Figure 8.4, examples are shown of these plants, installed and operational over 40 years.

There has been considerable discussion about the safety and reliability of nuclear power to date; however, the experience in UK has been one of steady expansion, learning and satisfactory operation.

The prize from nuclear power could be its potential for large quantities of energy without the necessity to burn reducing quantities of carbon sources, coal, oil or gas with their associated pollution.

Figure 8.4 Great Britain's operating nuclear power stations from 1956 to date. (a) World's first nuclear power station generating electricity, Calder Hall, 4× gas-cooled reactors and 4×60 MWe, 4-pole turbo-generators, opened by HM Queen Elizabeth II, operational 1956, now demolished. (b) Wylfa nuclear power station, 2× Magnox gas-cooled reactors and 4×334 MWe, 4-pole turbo-generators, commissioned 1971, now decommissioned. (c) Torness nuclear power station, 2×AGR gas-cooled reactors and 2×660 MWe, 2-pole turbo-generators, commissioned 1989, operational. (d) Sizewell B, 1×PWR water-cooled reactor and 2×600 MWe, 2-pole turbo-generators, commissioned 1995, operational.

However, nuclear power involves operating reactors containing high energy nuclear fuels, which in the event of an accident could release large quantities of poisonous radioactive materials into the Earth's atmosphere.

8.3.3 Evolution of nuclear power

The steady application of nuclear power and its evolution, in the United Kingdom, has been repeated in the former Soviet Union, USA, France, Japan, China and 25 other countries around the World.

An impression of that technological evolution from British nuclear reactors can be seen in Figure 8.1.

Early UK plants had their reactors, encased either in steel or steel-reinforced concrete enclosures, with double containment and the nuclear fuel enclosed within a core, using graphite as a moderator, coolant gases transferred heat from the enclosure containing the nuclear fuel to the steam-raising plant.

Examples of these nuclear plants, using CO_2 gas as the coolant, are shown in Figure 8.1(a)–(c).

These earlier reactors with low steam output pressures, for example at Calder Hall, were used to drive low speed 4-pole turbo-generators.

Whilst larger, higher pressure steam output reactors, for example at power stations like Torness, an advanced gas-cooled reactor (AGR), and Sizewell, a pressurised water reactor (PWR) using water as a moderator, were being used to drive larger faster 2-pole turbo-generators, Figure 8.1(c) and (d).

Meanwhile the Boiling Water Reactor (BWR) was being developed for electricity generation, a simpler and cheaper form of reactor than the PWR technology developed for nuclear submarines. The BWR design raises steam in the reactor directly for use in the turbines and has a circular condenser below and surrounding the reactor.

Both BWR and PWR technologies have been deployed in Japan, China and USA, whilst in France, PWR technology alone has been developed.

The United Kingdom continued first with the Magnox and then AGR gas-cooled technologies, before later turning to PWR technology.

Gradually, the PWR reactor has become more widely used for electricity generation, because of the designs' long development, security, safety and universal application, not only from 140 nuclear submarine propulsion applications, but also gradually in large, full-size, shore-based power stations.

Sizewell B was the first example in United Kingdom of a PWR-equipped power station to be built.

Other alternative modern reactors had also been developed:

The sealed Canadian Pressurized Heavy-Water Reactor (CANDU). The acronym refers to its use of deuterium oxide, heavy water as a moderator and the use of uranium fuel. CANDU reactors were first developed in the late 1950s and 1960;

The sealed Russian Water-Water Energetic Reactor or VVER is very similar to the PWR but with a higher output steam temperature, so that it can supply larger power output to high speed turbo-generators.

There has been international debate about the relative safety of different nuclear power plant designs, but to ensure ease of operation and reliable power generation, the following principles have provided a sound basis:

Enclosing the reactor and nuclear core within a sealed steel sphere or tube, based on the widespread safe operational experience from nuclear submarines;

Closed-circuit primary 3 or 4 pure water-cooling loops, transferring heat out of the nuclear core;

Heat then being transferred to a secondary boiler circuit for raising steam, exiting the reactor plant to power steam turbines but isolated from it, independent of the reactor core, the PWR principle. The deployment of duplicate redundant safety systems ensures simple and secure operation.

There has been considerable debate within and between countries about the use of nuclear power.

The United Kingdom embraced nuclear power early, initially developing the gas-cooled, graphite-moderated reactor inside a steel pressure vessel, using CO_2 as the coolant, with steel containment but subsequently, as power outputs increased, the steel containment was replaced by cheaper, thicker, steel-reinforced radiation absorbing concrete, with acceptable experience from both enclosures.

More recently, the UK deployed the spherical, sealed steel pressure vessel, PWR nuclear reactor, of Westinghouse design, in a power station at Sizewell B. The United Kingdom is currently building a new PWR power station, of EDF French design, at Hinkley Point C with another similar PWR planned for Sizewell C. In the United Kingdom, the nuclear industry has had a good safety record, despite the early incident at Windscale. The United Kingdom currently has nine operational power-producing nuclear reactors.

Some countries, such as Italy, have ceased all nuclear power operations because of the potential seismic risks in their peninsula.

Other countries, such as Germany which had nuclear power from the 1960s until being phased out in 2023, because of democratic opposition to nuclear power dependence. This has meant that Germany has shut down 27 nuclear power plants and is decommissioning them in Germany both East and West.

France has a very large, consistently reliable fleet of PWR nuclear power stations with 56 operable power-producing nuclear reactors, developed over 60 years, with a strong safety record.

The USA has invested in nuclear power generation, first using the cheaper, simpler BWR reactor and subsequently adopting the PWR, to power their plants, based on nuclear submarine technology, they currently have 92 power-producing nuclear reactors.

Russia, former Soviet Union, also invested in nuclear power generation, despite its considerable coal, oil, gas and hydro resources. It initially deployed lower cost, graphite-moderated, water-cooled reactors, with a single concrete enclosure, the RBMK reactor. More recently, it has developed the more powerful, higher pressure VVER reactor. Russia currently has 38 power-producing nuclear power reactors and has exported reactors to neighbouring Soviet countries and sold VVER plants to Germany, China, South Africa and Iran.

Japan invested in nuclear power, despite country's seismic activity, Figure 3.7, because of its lack of fossil fuels for power generation. Initially Japan deployed well-tried UK-style gas-cooled reactors and then switched to US-style BWR reactors of GE design with a good safety record until the accident due to the Fukushima tsunami, and then subsequently more modern PWR reactors. Japan currently has 33 operable nuclear reactors.

China has imported several well-tested nuclear reactor designs from abroad, including the BWR, PWR and VVER, building a large fleet of nuclear power stations to reduce dependence on polluting coal-fired power stations, using poor-quality local coal. China has learnt from growing nuclear operational experience and increased safety knowledge, having recently developed its own PWR design, the HuaLong reactor, and is now exporting these plants overseas to Pakistan and elsewhere. China currently has 51 operational nuclear power units.

The electrical energy delivered, to date, by all these nuclear operators is shown in Figure 8.5. The dip in growth of nuclear power development following 2011 was a consequence of the nuclear accident at Fukushima Daiichi Nuclear Power Plants in Japan in that year and subsequent plant shutdowns, see a following section for a description of that incident.

The largest reduction in output was in East Asia but following that date there has been a slow but steady increase in electrical output from nuclear power stations.

The author was able to witness, first-hand, some of this nuclear power growth in the following locations and years:

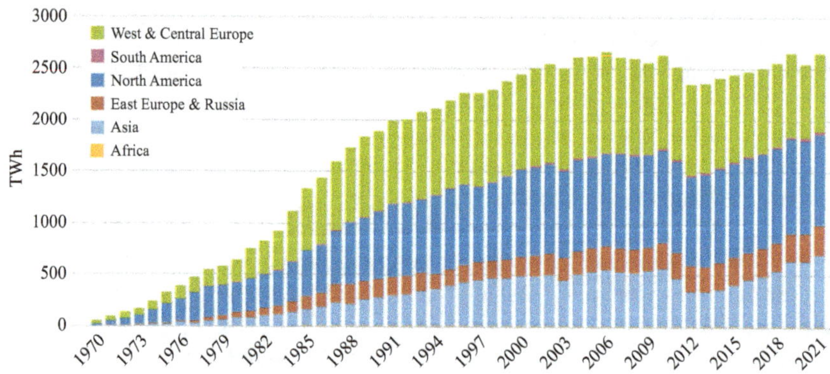

Figure 8.5 Electrical power delivered world-wide from nuclear power stations, operating from 1970 to date

- 1975, as a graduate student, whilst visiting CA Parsons Ltd factory in Newcastle, seeing 2 new 750 MW, 2-pole turbo-generators ready for despatch to the Bruce CANDU nuclear power station in Canada, subsequently commissioned in 1978 and 1979;
- 1981, as a CEGB engineer, attending the UK's older Dungeness A Magnox nuclear power station to resolve an explosion from the main bearings of one of its four 142 MW, 4-pole turbo-generators cooled by hydrogen, commissioned in 1965. The explosion was caused by the ignition of hydrogen escaping from the generator's bearings;
- In 2017, 42 years later, the author worked briefly in China as a consultant to Dong Fang Dian Ji (东风电机) with an electrical engineering factory DFEM in Deyang, Sichuan Province.

There he witnessed the growth of China's coal- and nuclear-fired electrical power generation, seeing the following generators under construction in the factory:

- 3×600 MW 2-pole turbo-generators being built for new coal-fired power stations;
- 2×1,000 MW 2-pole turbo-generators being prepared for delivery to the TianWan nuclear-fired Russian VVER reactor power station, commissioned in 2021;
- An 1,100 MW 4-pole, turbo-generator being prepared for delivery to China's new HuaLong PWR reactor power station entering commercial operation in 2023.

The author's nuclear vignettes, over that 56-year span, demonstrate the nuclear industry's steady but relatively rapid growth in terms of unit size, power, technology and location:

- From 1965 when two gas-cooled Magnox reactors at Dungeness A power station in UK powered four 142 MW 4-pole turbo-generators;
- To 1979, when two CANDU reactors at Bruce power station in Canada each powered a 750 MW 2-pole turbo-generator, see Figure 8.6;
- Then 2021, where four modern VVER-1200 reactors at TianWan power station in China (田湾核电站), each powering a 1,060 MW 2-pole turbo-generator, see Figure 8.7.

8.3.4 Nuclear accidents and lessons learnt

There have been concerns over the last 80 years about the relative safety of nuclear plants, compared to other thermal power generation units.

To place nuclear power in context, it is important to note that, world-wide, there have been four significant nuclear plant incidents, as follows refer to Figure 8.4 for images of these plants:

- UK, Windscale Nuclear Test Plant.
  10 October 1957;

Figure 8.6 *Bruce Nuclear Power Station Unit 1, Canada, operating from 1977 to date*

Figure 8.7 *TianWan Nuclear Power Plant (田湾核电站) Jiangsu Province, China, Units 1 and 2, each with 2 VVER-1000 reactors rated at 1,060 MWe*

Fire in a gas-cooled, nuclear test plant, designed to provide a reactor core for the irradiation of samples, having a single concrete enclosure with graphite moderation and an open-loop, air ventilation system;
This was not a plant for electrical generation, but for irradiating materials to be used in the nuclear industry;
The fire was extinguished with water within 24 h;
Loss of life none, radiation release and symptoms caused are uncertain, but one estimate proposed the accident could have initiated 100–240 long-term cancer fatalities.

- USA, Pennsylvania, Three Mile Island (TMI) Nuclear Power Station, with 2 operational Units, respectively powering 780 and 880 MW turbo-generators.
  28 March 1979;
  Unit 2's PWR experienced a cooling system malfunction causing a partial meltdown inside the reactor core. The reactor was shut down and the meltdown restricted to Unit 2's pressure vessel, from which, there was an emergency release of water–vapour/steam and other material. The interior of the core was badly damaged, and it never operated again.
  Because of health concerns, Pennsylvania's Department of Health kept a symptom register of more than 30,000 people living within 5 miles of TMI at the time of the accident and subsequently.
  This register was maintained for nearly 20 years until 1997, by when no evidence had been found of unusual health effects.
  Further epidemiology studies have not shown any increase in cancer because of the accident. However, almost $25 million was paid in insurance settlements by the power company to people, who then agreed not to discuss their injuries in ongoing litigation.
  Recent studies support a hypothesis that the TMI radiation doses did give rise to increased cancer incidence around TMI but numbers are uncertain.
- Ukraine, Soviet Union, Chernobyl Nuclear Power Station, with four operational units each unit powering two turbo-generators.
  26 April 1986.
  Unit 4's RBMK, High-Power Channel Reactor with a single concrete enclosure which contained a water-cooled, graphite moderated, reactor delivering steam to two 500 MW turbo-generators.
  Unit 4 exploded during a delayed power-down test on one of the reactor's two generators. The delayed test put the reactor into a condition where control rods had to be fully withdrawn to raise power, a situation that caused an explosion of the reactor and its concrete enclosure, exposing the reactor core. This in turn caught fire and ejected radioactive fuel into the external environment. The reactor and power station were totally destroyed.
  There is a consensus that approximately 30 people died in the days and months after the disaster, from immediate blast trauma and acute radiation. This number rose to 60 in total, due to the radiation spread by the explosion and the effects of fire and smoke from the site.

Fewer than 100 deaths in total were directly attributed to the accident, including later radiation-induced cancers in the decades since, but there have been varying reports, over subsequent years, of deaths further afield in Ukraine and Byelorussia.

However, there is considerable debate concerning the accuracy of projected numbers of deaths, due to long-term health effects, yet to occur.
Estimates ranged up to thousands for the most exposed people in Ukraine, Belarus and Russia.

At the first international conference on the Chernobyl disaster, the International Atomic Energy Authority (IAEA) established a figure of 4,000 deaths as the total number of projected deaths caused by the Chernobyl accident over the long term but did not make that figure official.

- Japan, Fukushima Daiichi Nuclear Power Plant, with 6 BWR power units.
16 March 2011

An offshore subsea seismic event at a tectonic plate separation off the East coast of the Japanese main island, the Tōhoku earthquake occurred on a fault-line, see Figure 3.7, initially damaging the power stations' substation switchgear and overhead lines at the Fukushima Daiichi nuclear power complex. This initiated an emergency shutdown of 3 of the operational BWR reactors located on-site. Reactors 4 and 5 had been shut-down before the incident, while the sixth reactor had not yet gone on load.

A tsunami was generated by the severe offshore seismic event, raising a tidal wave coming ashore and overtopping the protective concrete barrages surrounding the Fukushima Power Station Units.

This tidal wave flooded the various unit basements, interrupting power generation and reactor services, cutting-off the electricity supply, emergency cooling of 4 of the reactors and causing a loss of power, light, and instrumentation in their Control Rooms, before very limited emergency arrangements could be made. Each Unit's control rods were correctly raised into their BWR reactors to terminate the nuclear reactions. But decay heat from the fuel continued to raise water temperature and pressure in each reactor, while hydrogen gas started forming around the control rods in the water because of their zircalloy coatings.

The effects of the tsunami within all the Fukushima Units seriously delayed the initiation of rescue, remedial and repair work, hindering the pumping of water and restoration of electrical services.

The three damaged reactor buildings were Units 1–3.

A subsequent series of consequential hydrogen-air explosions occurred within the reactor buildings of Units 1–4, with water–vapour/steam and other material being ejected, causing further internal reactor structural damage. Some timely and heroic venting of water–vapour/steam by station staff prevented a similar explosion in Unit 2.

There is no doubt that the sudden initial effect of the Tsunami and possible successive inundations prevented station's staff from undertaking the immediate remedial actions that they wanted. However, leadership, growing cohesion and the arrival of electrical, fire, military, plant service assistance and overseas help gradually paid-off and the radioactive impact of the accident was reduced.

Estimates of radioactivity released, during the incident from Fukushima's three active units ranged from 10% to 40% of that from Chernobyl's single unit, while the contamination in the areas around Fukushima by the accident are estimated to be 10%–12% of those from Chernobyl.

As of September 2018, two nuclear station workers had died as a result of the accident but many had received low level radiation effects, some significantly shortening life.

The maximum predicted eventual cancer mortality and morbidity estimates are 1,500 and 1,800, respectively, according to linear no-threshold theory.

However, there were no other deaths from radiation exposure in the immediate aftermath of the incident, although around 1,600 deaths, non-radiation related, occurred during Fukushima's nearby population evacuation.

In total approximately 18,500 people died due to the earthquake and tsunami, more than 2,500 people are still officially reported missing and a further 6,000 suffered injuries.

Rama C. Hoetzlein has produced a valuable post-Fukushima graphic in Wikipedia, Figure 8.5, that compares the impacts of the above four nuclear incidents, showing that

- Chernobyl (1986) had the worst impact, with a radiation release level more than 100,000 mSv/h;
- Fukushima (2011) had the next impact, with a radiation release level up to 100 mSv/h, see Figure 8.8 for a comparison between Fukushima and other nuclear accidents;
- Three Mile Island (1979) had a much lower impact, with radiation release level at 0.0001–0.001 mSv/h;
- Windscale (1957) did not register on the above scales but that could be because accurate measurements were unavailable at that date.

Engineering has always involved risks of failure, whether from:

- In the nineteenth century, exploding steam locomotives or appalling train crashes;
- In the twentieth century, the early unreliability and crashes of motor vehicles;
- In the second half of the twentieth century, recurrent but steadily diminishing aviation accidents.

As a result of each case above, technology has developed, driven by safety, risk, cost and overall benefit.

- Since the cases above nuclear accident rate figures have steadily declined to relatively low figures in the power generation industry so that fourth and fifth generation reactors look likely to reduce these accident rates even further.

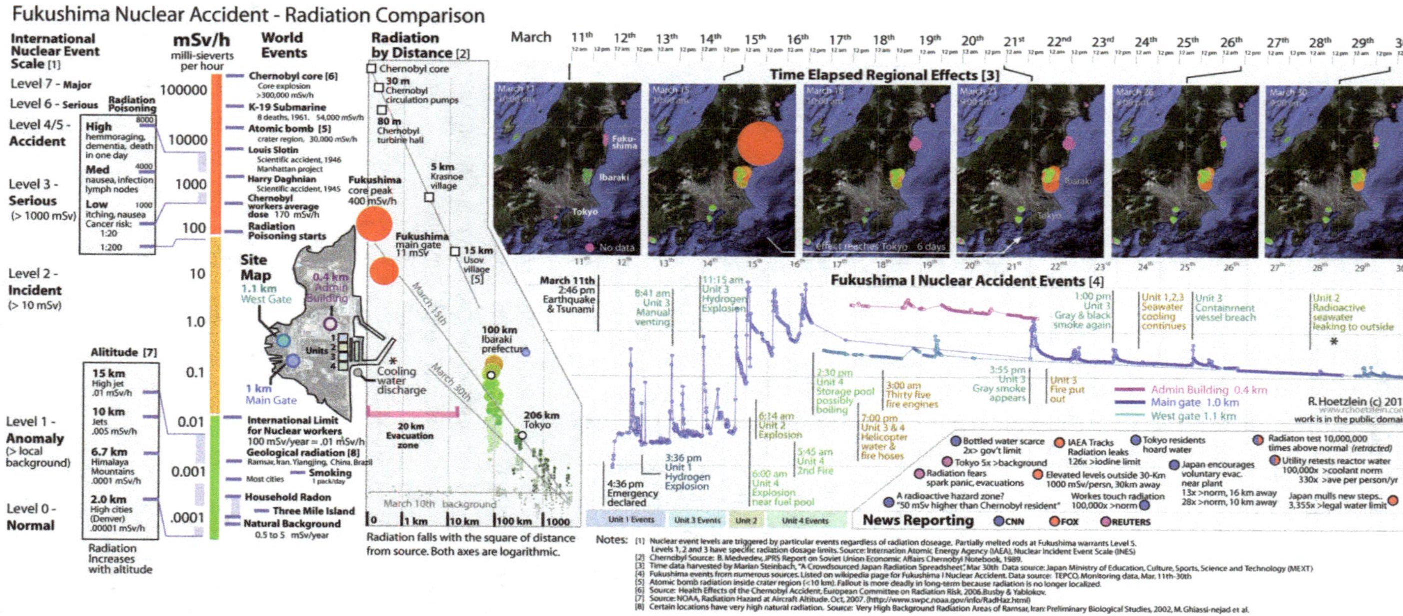

Figure 8.8 Comparison of the effects of Nuclear Accidents to date. Rama C. Hoetzlein, own work, released in public domain.

- Nuclear power has had a similar experience, compressed into a shorter period of time, 80 years from the second half of the twentieth century to start of the twenty-first century, but with higher stakes, because of the potential risks of contamination and pollution from larger plant sizes.
- The resolution of such difficulties in any technology is information, knowledge and understanding.
- Hoetzlein's diagram, Figure 8.5, uses the Fukushima accident as an example to compare nuclear power accidents.
- The route to clean energy will be marked by the steady succession of innovations, failures, their resolution of the myriad difficulties and then success for the acceptable technologies.

8.4 Energy storage for conventional plant

8.4.1 Historical

As a result of the technological advancements described in previous sections, there has been increasing interest in methods for storing energy, so that when demand is low generation, plants can continue and energy be stored for later use.

Early high-power energy sources, such as coal and oil, did not lend themselves readily to storage techniques, as coal production involved continuous transportation, generally by rail, of the fuel from mine to power station, whereas as oil eventually began to rely on oil pipelines and sea-tankers to deliver the product to its customers.

However, in both the coal and oil industries, there has been considerable ingenuity to achieve some sort of buffer-storage to ensure security if fuel supplies to customers. For example,

- Coal producers retained coal stock-piles at the mines.
  Whilst the users, steel mills and power stations, developed sophisticated coal heaps, conveyors and record systems on their sites, to be able to maintain some level of supply when mine or rail strike problems occurred.
- Oil producers also retained some oil tank storage at their shipping delivery ports.
  They also employed fleets of oil tankers, constituting an effective mobile store. However, consumers had oil tanks on their sites as a reserve.
  The use of interconnected computing has allowed users to retain a reasonable level of control of these types of energy stores.
  However, these were only adequate under relatively stable normal conditions. Wars and local disasters could readily disrupt these preparations.

8.4.2 More diverse storage mechanisms

8.4.2.1 Traditional hydro schemes

Hydro-electric schemes have an intrinsic storage capability since water can be released through the hydro-electric machinery whenever electrical energy is needed.

However, dams also have a requirement to maintain water flows and there can be conflicts between the need to maintain river flows, the electrical demand and the desire to retain some water head for energy storage needs. Large hydro-electric schemes have plans in place to balance these conflicting requirements, and hydro-electric schemes can be of considerable help to balancing a large electrical Grid, for example in Canada, the USA, France and China.

That task is more difficult in smaller countries like UK with relatively small hydro-electric resources.

8.4.2.2 Pumped storage schemes

A development of hydro schemes, becoming more visible since the growth of demand for variable renewable energy, has been the purpose-designed pumped storage scheme, where a hydro-electric scheme is able to both generate electricity and reverse to pump water up to a reservoir as a form of storage.

These schemes are technologically feasible in countries with a mountainous or hilly terrain but their civil engineering costs are high and construction times long.

However, experience from hydro-electric schemes has been that such projects can have a long and useful operational life, provided that they are well-maintained.

Detailed examples of pump storage schemes in operation are given in Chapter 9.

8.4.2.3 Compressed gases

Some engineers have advocated the development of compressed gas machinery connected to high pressure vessels or underground stores. Such arrangements would be valuable if they are to be powered by variable renewable sources such as wind turbines or wave machines. They could provide the means to store the variable energy produced by wind or wave farms.

There have been some examples of wind turbines driving air compressors or wave machines compressing air but as yet these technologies have not been fully industrialised.

8.4.2.4 Flywheels

Flywheel-storage power systems use the inertial energy of a spinning flywheel to generate electrical power in a generator to inject electrical energy into a grid at times of falling frequency.

Such flywheels are a comparatively small storage facility but with a peak power output up to 20 MW.

Typically, they are used to stabilise electrical power grids with rapidly varying loads, such as electric subway train systems. They can also serve as a short-term compensatory storage mechanism.

Compared to other plants, such as pumped storage, with capacities up to 1,000 MWh, the benefit of flywheel storage is that it can supply electrical energy in the range from kWh to a few MWh.

A flywheel-storage system is comparable to battery storage power plants.

Possible areas of application are places where electrical energy can be obtained and stored and must be supplied again to compensate for example, fluctuations in the seconds range in wind or solar power. These storage facilities consist of individual flywheels in a modular design. Energy up to 150 kWh can be absorbed or released per flywheel. Through combinations of several such flywheel accumulators, which are individually housed in buried underground vacuum tanks, a total power of up to several tens of MWh can be achieved. The electrical connections power LV motors via a DC intermediate circuit and the power converter systems are comparable to those found in plants used in the high-voltage direct current transmissions application.

Sometimes battery storage power stations are built with flywheel storage power systems in order to conserve battery power. Flywheels can handle rapid fluctuations better.

8.4.2.5 Flow batteries

There has been growing interest in the use of sophisticated batteries to provide support for the electrical grid during transient load variations.

Invinity Energy Systems and Pivot Power, both parts of EDF Renewables, have energised a large flow battery as part of the Oxford Energy Superhub (ESO). This flow battery pioneers an integrated approach to decarbonising power, transport and heat. The 5 MWh flow battery system manufactured by Invinity is combined with a 50 MWh Wärtsilä lithium-ion battery to operate as a single AC grid energy store.

This is the largest directly transmission-connected battery installed in the United Kingdom and the largest vanadium flow and lithium-ion hybrid batteries deployed to date. The hybrid approach benefits from the strengths of these two battery technologies, increasing grid resilience and providing a flexible system for supporting the penetration of more renewable generation into the UK grid.

In the ESO system, the vanadium flow batteries front-end the storage asset, acting as a first line of response when the flow battery is called into service. Once the response exceeds the capacity of the vanadium flow batteries, the lithium-ion battery starts operation.

The invinity flow battery does not degrade with use and can cycle indefinitely, and it cushions the system grid while reducing lithium-ion battery wear.

8.4.3 Importance of storage for balancing networks

Large extended electrical power networks can now deliver power over enormous distances. However, they have another valuable capability that they can transfer energy from locations that are energy rich to locations that are energy poor. In addition, there maybe energy storage mechanisms within an electrical grid, capable of either slow or fast energy input, for example:

- Low capacity but reliable stores, such as high-speed rotating generator shafts with small but fast response rotational kinetic energy. These are controllable by the fast reaction generator AVR, valuable for balancing the power outputs between closely interconnected machines;

- Medium capacity stores, such as small water reservoirs or batteries that can be pumped full and discharged at times of energy need to balance the transmission grid;
- Large capacity high level water reservoirs that can release energy from the less-populous distant mountain regions to more populous cities with high energy demand. These are controllable by the long-distance HVAC and HVDC transmission Grid interconnectors.

This ability to store and transfer energy within the electrical grid is where value exists in international energy networks, since such techniques apply to other Energy Vector networks than electricity.

8.5 New network examples

The following sections will describe, not exhaustively, a number of different energy and transport networks, from source to user, demonstrating the World's growing human energy interconnectedness.

These networks started with terrestrial trade, for example overland from China to the West, expanding to maritime coastal trade, followed by world-scale maritime trade routes then land-based rail and road networks followed most recently by aviation routes.

These trading routes and energy transport networks now depend more heavily on modern communication and computing networks.

8.5.1 Oil and natural gas pipelines

As the oil and gas industries expanded in the twentieth century, pipelines were installed onshore to distribute their output.

Today, one of the world's longest crude oil pipelines runs from Edmonton, Canada, to Chicago, USA, then back to Montreal, Canada runs about 2,353 miles with 82 pumping stations, maintaining a flow of over 1.6 million barrels of oil per day.

These networks have expanded steadily from the 1930s to the present day and include the development of offshore network of pipes, for example in the North Sea oil fields.

This expansion of transmission of oil by pipeline has produced new specialist companies. In United Kingdom what was formerly the British Pipeline Association is now BPA, owned by the UK's largest developers, BP and Shell. Figure 8.9 shows Great Britain's pipeline network which pumps oil, aviation fuel and other flammable liquids around the country. Its network is based upon the pipelines laid in Great Britain during the Second World War, Figure 6.2.

Figure 8.10 shows the larger European pipeline network.

Since the 1970s, the natural gas industry has also expanded world-wide.

Their present-day network in Europe has steadily expanded, as shown in Figure 8.12.

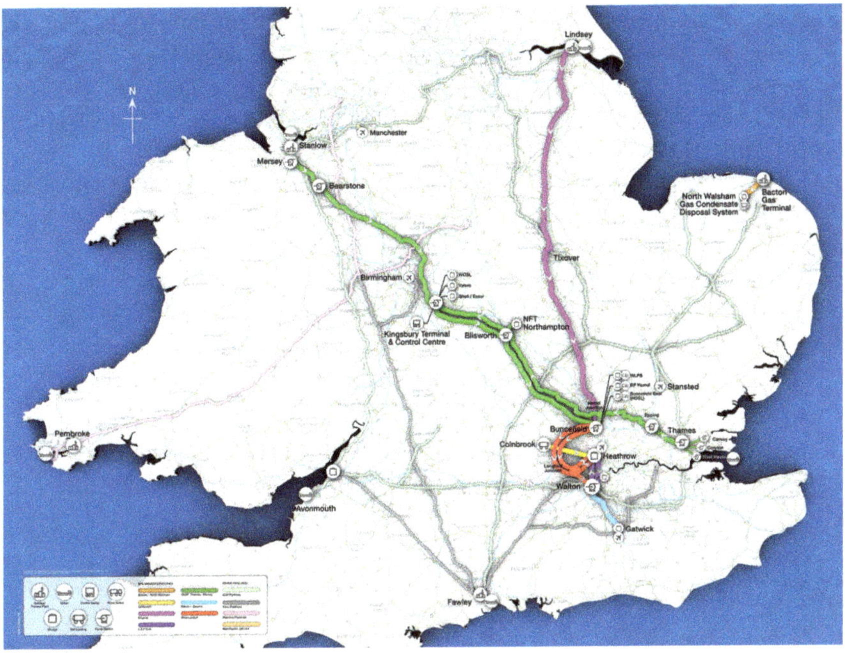

Figure 8.9 Oil pipeline network, Great Britain British Pipeline Agency, BPA, 2021

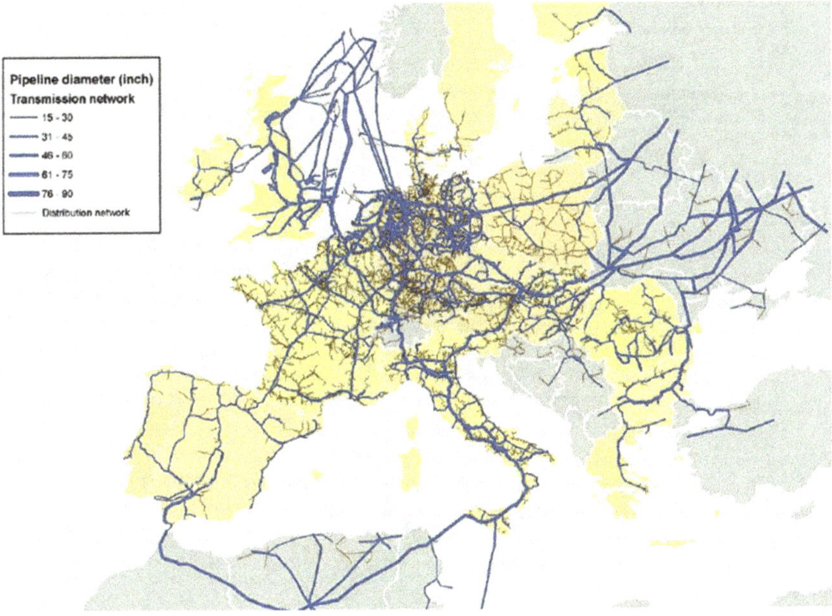

Figure 8.10 Oil pipeline network, Europe, 2021

This network now includes pipelines, Nord Zee 1 and 2 from Russia, although at the time of writing this book, these are disconnected because of the Russia–Ukraine war.

The flexibility of gas, its easy transportation by pipeline and the development, after the birth of the jet aircraft and of the more thermodynamically efficient gas turbines for power delivery, has meant that later in the twentieth century natural gas has become an important Energy Vector as a fuel for gas turbine-driven power stations and chemical plants.

Figure 8.11 is a valuable graphic showing the present natural gas network in Europe with indications of main product flows. The figure also contains useful information about:

- The spread of fuels used in Europe, in percentages, including oil, gas, nuclear, coal, renewables and hydroelectricity;
- Imports of gas from various overseas countries, including the Americas, Africa and the Gulf;

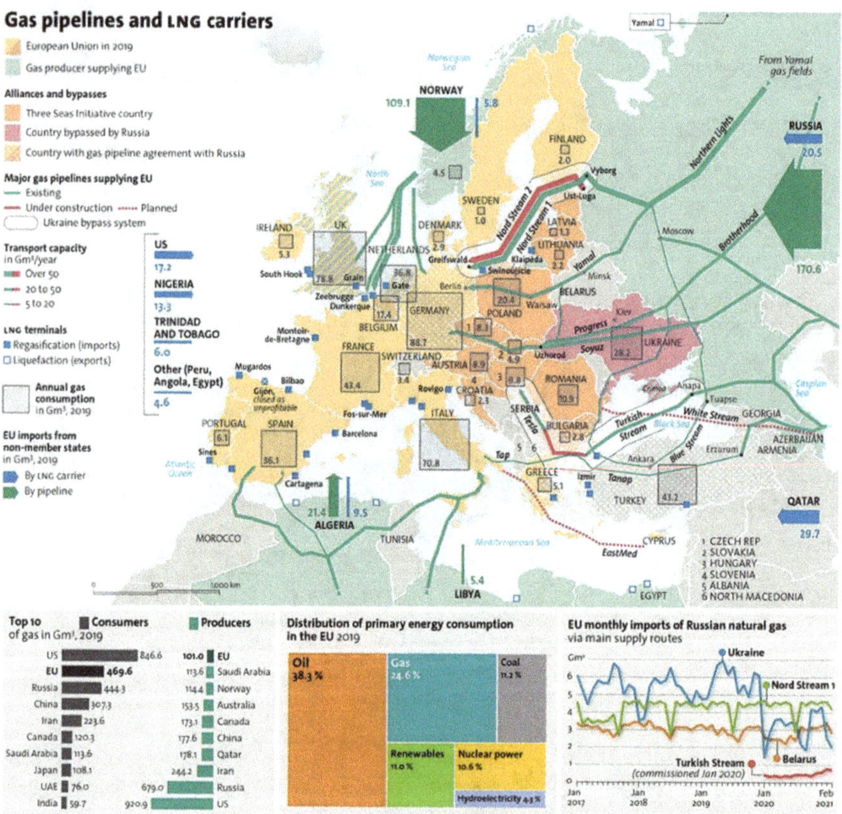

Figure 8.11 Gas pipeline network, Europe, 2020

- Graph of gas imports from Russia, Ukraine and Belarus up to the time of the Russia–Ukraine war;
- A useful table of producers and consumers.

The result of this expansion in the transmission of gas by pipeline has stimulated new specialist pipeline companies, for example Enagás (Spain), Fluxys (Belgium), GRTgaz (France) and SNAM (Italy).

The following gives a brief overview of the extensiveness of the current largest natural gas pipelines world-wide:

- West-East Natural Gas Pipeline in China, Xinjiang to Shanghai (上海), length 8,706 km;
- GASUN Natural Gas Pipeline, planned in South America, Bolivia to Brazil, length 4,989 km;
- Yamal-Europe Natural Gas Pipeline, Siberia Russia to Germany, length 4,197 km;
- Trans-Saharan Oil Pipeline, planned, Africa, Nigeria to Algeria, length 4,128 km;
- TransCanada Natural Gas Pipeline, Canada, Alberta to Quebec, length 3,227 km;
- Rockies Express Natural Gas Pipeline, USA, Colorado to Ohio, length 2,700 km;
- Transcontinental Natural Gas Pipeline, USA, Texas to New York, length 2,689 km;
- Trans-Mediterranean Natural Gas Pipeline, Mediterranean, Algeria to Italy, length 2,591 km;
- Northern Border Natural Gas Pipeline, N America, Canada to Chicago. length 2,239 km;
- Nord Stream Natural Gas Pipeline, Europe, Russia to Germany, length 1,221 km.

8.5.2 World electrical networks

Modern EHV electrical transmission networks have proved essential for transmitting electrical energy, particularly in large, developed countries. The following figures demonstrate this as follows:

- Figure 8.12 for the United States and Europe;
- Figure 8.13 for In China and Japan;
- Figure 8.14 for Africa and India.

These figures demonstrate the modern extent of electrical transmission.

8.5.3 Comparison of electrical grids with oil and gas pipelines

In general, electrical grids are nationally owned, whereas oil and gas pipelines are privately owned by oil and gas companies, who require governmental approval to lay those pipes.

The result is that electrical networks have grown extensively to meet the needs of industry, governments, and the population, covering whole countries.

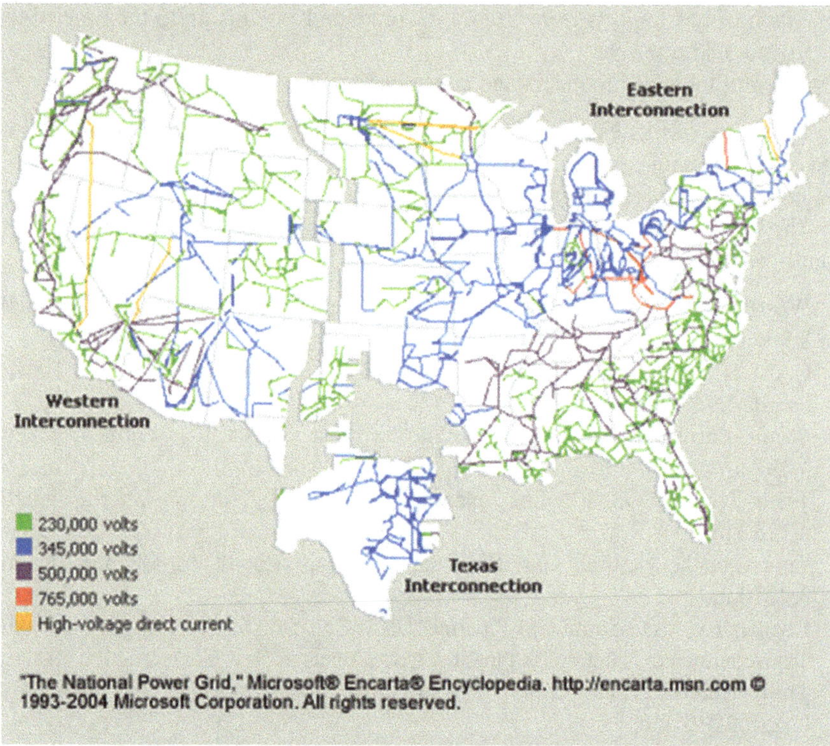

(a)

(b)

Figure 8.12 Showing highly developed electrical EHV networks in USA and Europe. (a) United States (b) Europe, Russia and the Near East.

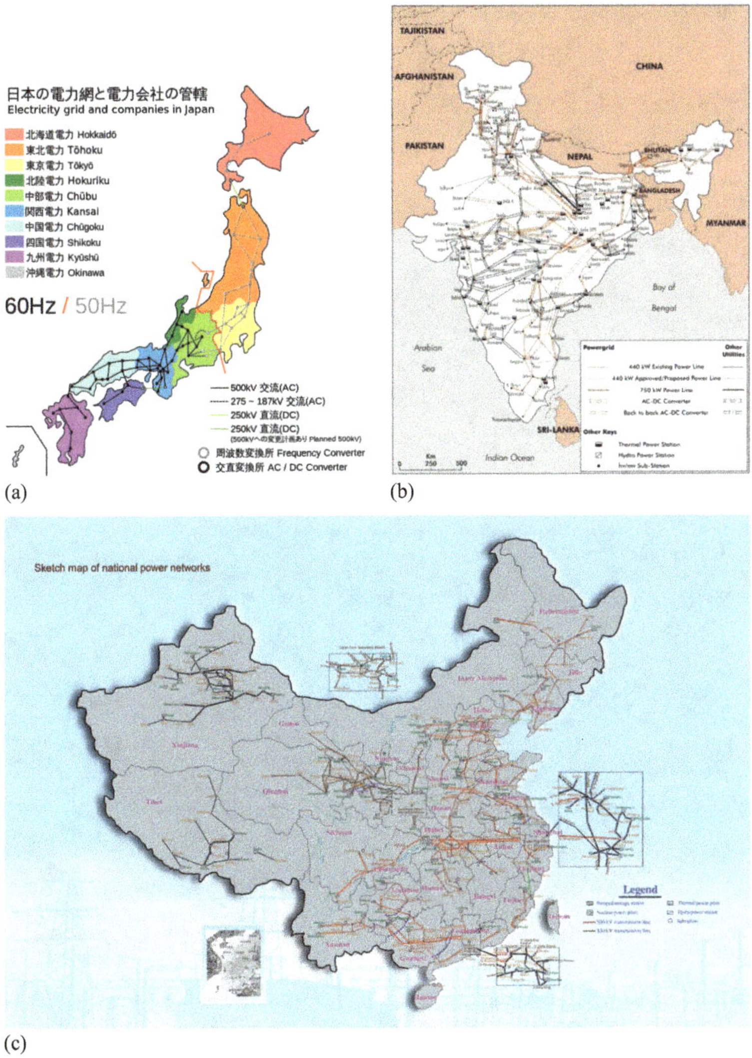

Figure 8.13 Examples of three highly developed EHV electrical networks in South and East Asia. (a) Japan, (b) India, (c) China.

Whereas oil and gas pipelines are much more closely controlled by the financial objectives of the oil and gas companies and are therefore less accessible than electrical grids.

However, the density of electrical grids, from LV, through to HV and EHV means that grid assets can become cumbersome and inflexible, through their size, extent and number of users.

This means that electrical grid developments can be limited by conflicting state and company energy requirements.

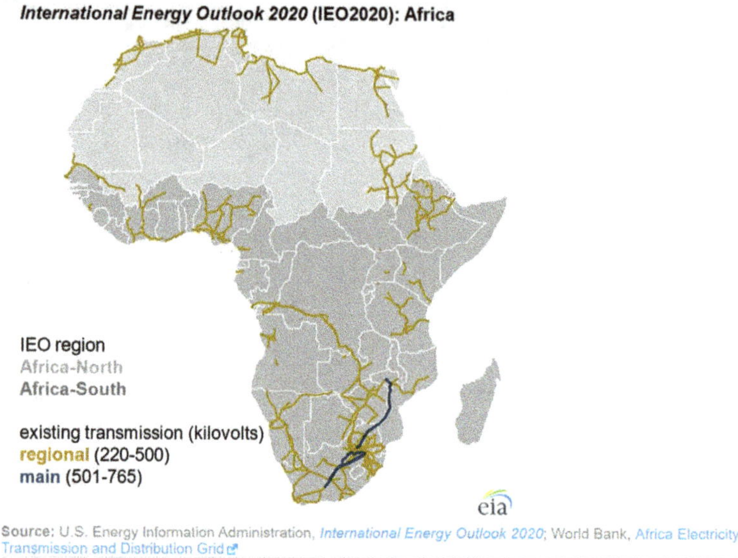

(a)

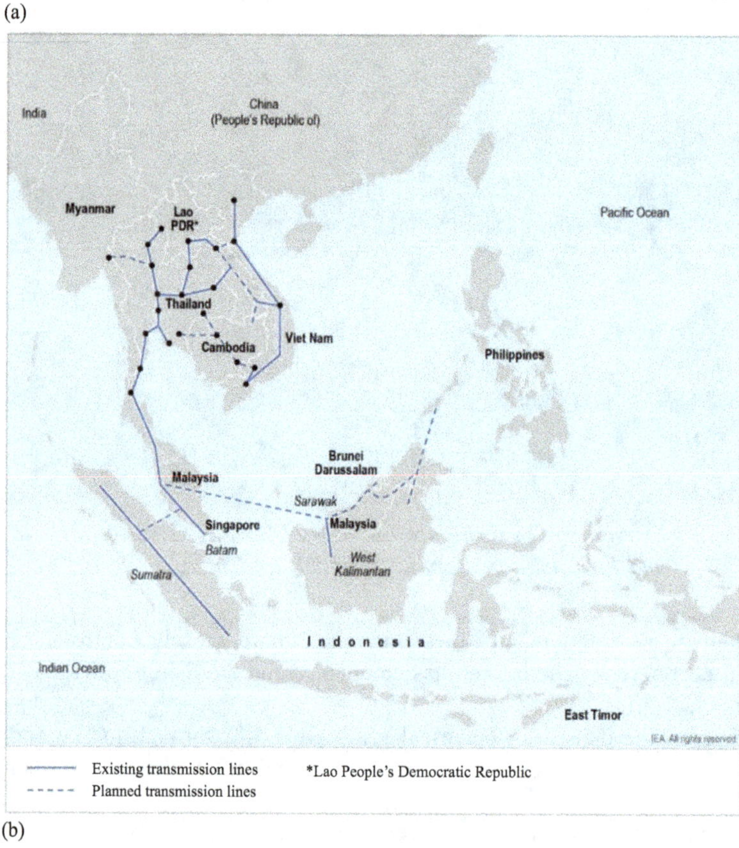

(b)

Figure 8.14 Examples of two less-developed EHV electrical networks. (a) Africa and (b) Southeast Asia.

This has recently become apparent in Europe, because of the large expansion in demand for renewable generation sources, particularly from wind, hydro and photovoltaics.

As a result, some European countries are trying to modify their rules for grid development, but this can be opposed by democratic opposition.

For example, at present in the United Kingdom in East Anglia, there are oppositions to the building of transmission lines to bring offshore wind energy from the coast to towns inland. There is strangely also opposition to such energy being transported inland by buried cables. Such opposition seems frequently ill-informed, quasi-luddite or at best mildly stupid.

On the other hand, in China, there has been energetic expansion of the grid, some of which has sometimes been opposed by the more conservative State Grid, responsible for these connections.

A different situation occurs in developing countries, for example in some parts of Africa or Southeast Asia or in remote island communities, where the lack of state capital can curtail the development of the Grid.

This was particularly noticeable to the author when working in Nigeria, where the expanding economy needed a strong Grid but limitations of Government funding and the country's huge size constrained the pressing need for Grid expansion.

Perhaps, in such cases, less centrally controlled Grid authorities, with access to a range of investment funds, could be beneficial.

8.5.4 Shanghai's urban rail network

In Section 6.4, the expansion of the London Underground railway system was described over a period of 120 years from 1880 to 2000 to over 5 million people.

When considering the important issue of networks and their importance, in this section, I must bring this up-to-date by shows the expansion to date of a broadly similar, but more complex Underground system than that of London. That is the metro network in Shanghai (上海), Figure 8.15, built over a much shorter period of 26 years, from 1993 to 2019.

The Shanghai Metro is an above-ground system, because of the city's soil conditions, making underground tunnelling difficult, and uses multiple-unit electric trains operating from 1,500 V DC overhead wires.

Refer to Duffy [16] for a good description of how modern integrated electrical mass-transit railways use electricity to power the trains and advanced electrical relay circuitry to locate and control large numbers of electrical multiple units on such a large urban system.

The author remembers visiting Shanghai in 1999, when the first three lines of the Mass Transit Rail system had been built in preparation for the Shanghai International Film Festival.

Figure 8.13 shows the present extent of the network, only 26 years later, with more than 21 lines, demonstrating the increasing transport and energy demands of this large dynamic trading city.

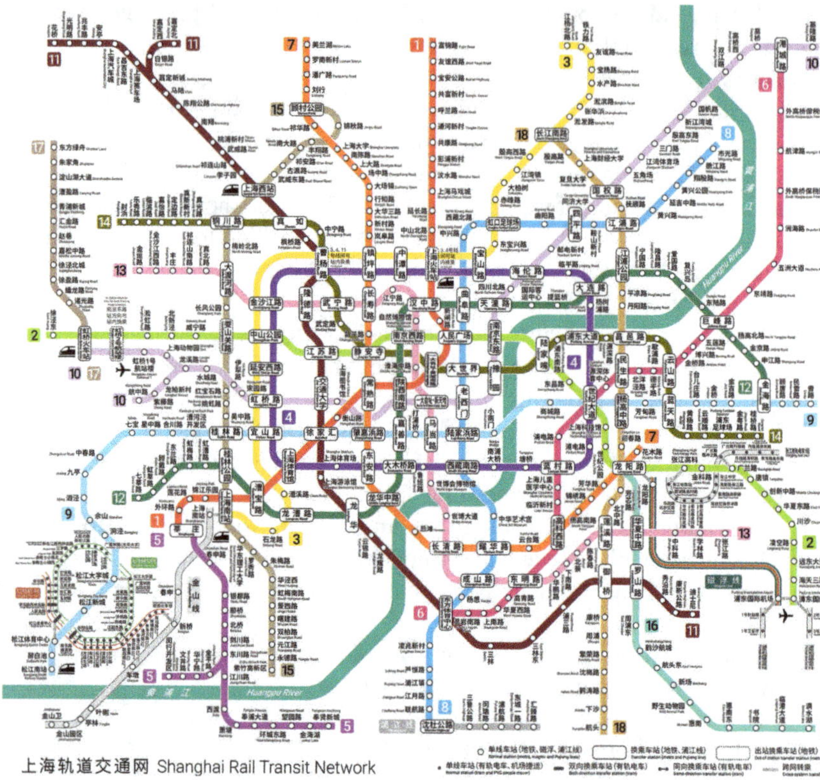

Figure 8.15 Extent of the Shanghai (上海) Mass Transit Rail system, 2020

8.6 Summary

The purpose of this chapter was to demonstrate the importance of Energy Vectors, whether for coal, oil, gas, electricity and their consequent networks.

These networks enable the connectivity of existing energy systems with consumers and the vital coal, gas, oil and electrical interconnections that they first initiated and now sustain.

The importance of the Energy Vector concept is that networks created to transmit energy at one period can evolve to transmit energy over longer distances and be modified or redesigned to carry other Energy Vectors.

So, the following steady but progressive changes and improvements in availability has been described:

- Coal by road then by rail;
- Oil by road then by rail and now by pipe;
- Gas by road then by rail and now by pipe;
- Coal to oil;

- Oil to gas;
- Coal, oil and gas to electricity.

As a result of these changes, first local city and industrial infrastructures were created. This was followed by wider infrastructures being developed to cover a whole country. Then these networks expanded to include adjacent countries, until whole continents are becoming included in comprehensive energy infrastructure systems, whether this is for coal, oil, gas, water or electrical energy sources.

Chapter 9
Cleaner energy technology examples – present

9.1 Introduction

The growth of electrical energy production and usage has been prodigious in size and extensive in technology, ranging in 1900 from thermal power generation units of a few hundred kW with efficiencies less than 5%, to in 2020 enormous thermal or hydro power units, in large generating stations each producing up to 3 GW of power, with efficiencies ranging from 30% to 65%.

From this 100-year expansion of scientific and engineering activity in power generation, this chapter gives examples of compact, modern power plants, showing positive progress in improving higher energy efficiency and lower environmental impact.

These developments are dependent not only on fuel and environmental conversion but also on computing and communications, making large-scale distribution possible, placing electrical technology, telecommunications and computing at the heart of future human energy usage.

The following technical descriptions briefly summarise a few of these recent developments in energy transformation for human use and the way in which we can hope for these developments to help us to turn the corner of climate change problems.

9.2 Lessons from fossil fuel applications

9.2.1 Modern turbines

With the progressive development of modern power technologies, we are starting to see much more efficient technologies.

Figure 9.1 shows a compact modern steam turbine with a thermal efficiency of 40%. delivering 200 MW from an AC generator.

This design benefits from the thermodynamic, turbine blade and electrical machine design improvements of the last 100 years.

The fuel for raising steam could be coal, oil, gas, wood-chip or industrial waste but the main conversion from heat to steam has been optimised in this arrangement from the design and operational knowledge of the last 100 years. An example of the deployment of this type of heat engine is shown in the following paragraph.

Figure 9.2 also shows the dramatic changes in steam and gas turbine blade designs, over the last 120 years, necessary to achieve modern power and aero turbine performance requirements.

Figure 9.1 Modern steam turbine generator, 2020, efficiency 40% left, successive HP and IP steam turbines, centre, two back-to-back LP steam turbines, right, AC turbo-generator, 200 MW, 50/60 Hz, 3 phase

9.2.2 Modern coal-fired power station with carbon conversion

From the late nineteenth century coal-fired, steam-raising plants powered first DC and then AC generators to deliver electricity.

The cost of the electricity generated depended upon the up-front capital cost of the plant and the running cost of the fuel, usually coal. The commercial mantra for lowering the cost of the electrical power generated was to increase the size and efficiency of the steam-raising plant and lower the coal supply cost.

The example described here Drax power station in the United Kingdom, based on UK thermal power station experience from 1900 to 1970, carrying out many improvements, achieving the thermodynamic pinnacle in a large, concentrated, advanced, fossil fuel fired, electricity generation plant, at that time one of Europe's or the World's largest coal-fired power stations.

These lessons have also been learnt and applied around the world.

The significances of the changes at Drax power station are that they represent the start of a transition from traditional concentrated fossil-fired energy, towards more environmentally sensitive approaches to electricity generation, using more complex plants to mitigate the polluting effects of fossil fuels and the early application of biomass as opposed to fossil fuels.

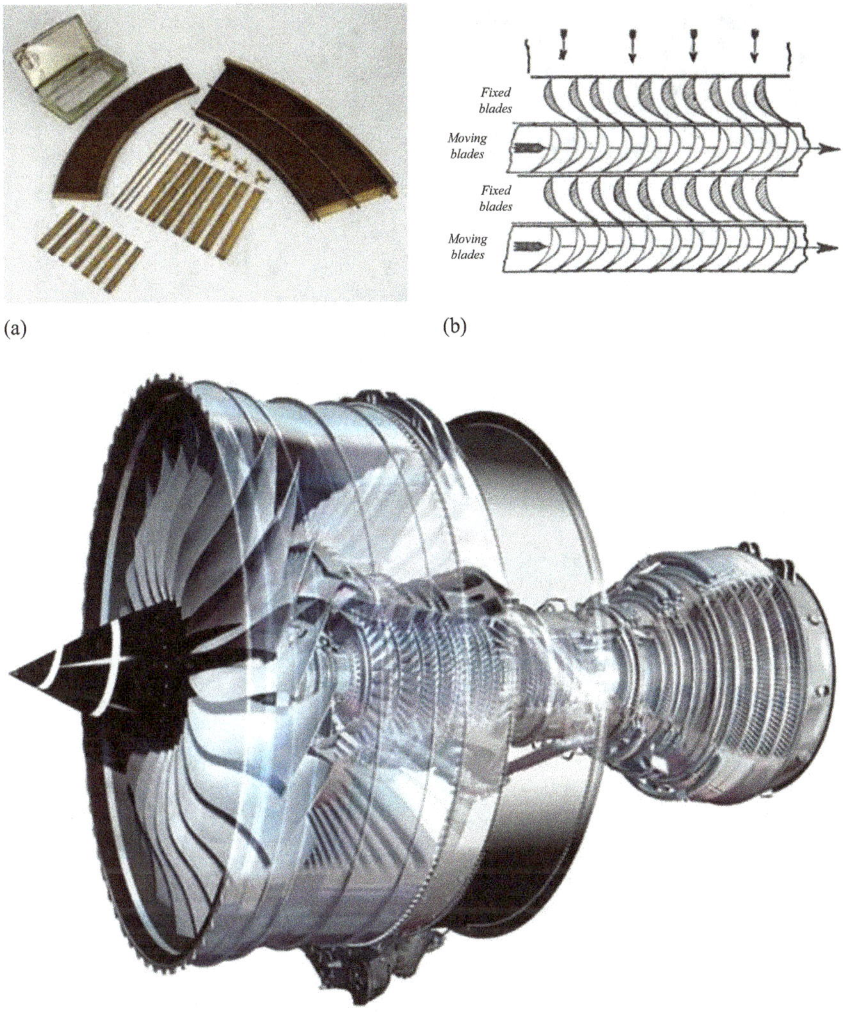

(c)

Figure 9.2 *Changes in turbine blade designs, 1890–2021. (a) Simple early Parsons flat blade steam reaction turbine design, 1890; (b) a simple steam reaction turbine, rotating and fixed blades, 1900; (c) Modern Rolls Royce Trent 3-stage aero gas turbine, compressor, power and thrust, showing modern variety of complex blade designs, 2010.*

Drax power station was the last and largest power station to be built in the United Kingdom, coal-fired or otherwise, with six off 660 MW steam turbine generators, and a final output rating of 3.9 GW. The power station was constructed in two phases, the first phase, see Figure 9.3(a), connected to the Grid in 1974 and the second phase connected in 1986.

(a)

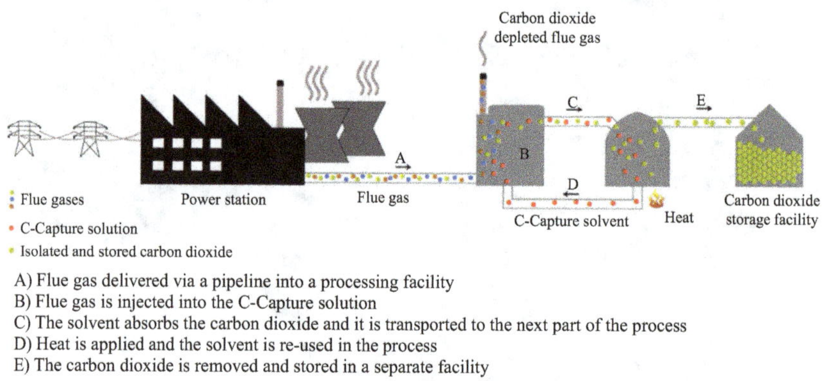

(b)

Figure 9.3 Drax Power Station, United Kingdom. (a) Turbine Hall at Drax Power Station, view of C A Parsons' 3 off 660 MW generators, driven by 3 stage (HP, IP and LP) 4-cylinder, impulse and reaction steam turbines, in the first phase. (b) A schematic diagram of the pilot scheme for Drax capturing and storing CO_2 emissions.

As one of the largest coal-fired power stations in the world at the time, with a thermal efficiency originally close to 40%, it benefitted from many thermodynamic and machinery lessons learnt in coal-fired, steam-raising electrical generation plants over the previous 50 years.

The coal supplies for Drax originally came from a highly mechanised super-pit nearby, which started operations in 1978, extracting coal from more than five mines, whose outputs were centralised by huge underground and overground conveyor belts to a rail site near Selby, supplying coal to the power station using continuous roundabout trains.

After a period of coal industry privatisation in United Kingdom, cheaper coal supplies became available from overseas and the Selby coal field declined, being closed in 2004. The closeness of these dates demonstrates the rapid changes in United Kingdom taking place in coal exploitation and power generation.

From 1982, there had been increasing concern about air-quality pollution effects from Drax power station's exhaust, particularly in Northern European countries, downstream from Drax's North Easterly chimney plume.

This was a collective problem for the whole fleet of United Kingdom's fossil-fired power stations. Therefore, from 1985, it was decided to install at Drax a Flue Gas Desulphurisation (FGD) plant, to extract CO_2 from the boiler exhaust and use chemical processes on the exhaust stream for the manufacture of gypsum-based products, see Figure 9.1(b). This extraction was initially applied to the manufacture of insulation board.

The FGD plant was installed between 1988 and 1996.

These modifications reduced overall plant efficiency below 38%, due to process power, abstracted from the power station's electrical output, and the flue back-pressure caused by the extraction plant. However, the modifications were environmentally successful, and such plants were subsequently applied to several other British coal-fired power stations, to reduce air-quality pollution.

From these lessons, Drax's steam turbine efficiencies were then progressively raised, to 39%, by improved design of turbine blade shapes.

Following further environmental concerns, National actions increased to reduce fossil fuel burning at Drax. Therefore, the plant was modified again to reduce its coal dependence and switch to biomass burning, using mainly woodchip imported from North America.

Currently 4 out of 6 of the generator units can operate with biomass and 2 with coal, now imported from abroad.

The CO_2 extraction process continues in operation, but extracted gypsum is now being used for a wider range of products than the originally planned insulation board.

However, one environmentally and commercially unsatisfactory aspect of the plant is that residual heat energy from the turbine energy extraction process is merely dumped into the atmosphere as cooling tower vapour and low temperature hot water into the river Humber.

The lessons from Drax, for all coal-fired power stations, are that, whilst we have been able to improve steam plant thermodynamic efficiency and reduce pollution from such a conventional fossil-fuelled power station, this could only be achieved by compromise. The overall solution should be to move away from fossil fuels.

However, the processes may initially be more expensive than conventional plant, moving electricity generation towards more complex, purpose-designed, environment-friendly plants.

These will then need to have a long enough life to defray the increased capital costs of these improvements over a reasonably short period. China still has abundant sources of coal, Figure 9.4, although that coal is largely low-grade lignite or sub-bituminous, rather than the more energy rich bituminous and anthracite, which occurred in UK. United Kingdom had, and still has, large reserves of bituminous and particularly anthracite coal, which we have stopped exploiting to reduce climate change effects; however, these resources could still be available in future for chemical purposes rather than simply raising steam.

But China is now the World's largest national exploiter and consumer of coal, Figure 3.1, using not only national resources but also importing coal from abroad, e.g. from Australia and the Gulf.

China is also the largest user of coal-generated electricity in the World with more than a thousand coal-fired power stations, see Figure 9.5 for the largest at Datang Tuoketuo, but then of course China is a very large, populous country.

9.2.3 Combined cycle gas turbine power stations

In the early 1970s, there was a move, initially from the United States, to apply gas turbine driven generators to industrial and domestic electricity production. This technology initially arose from advances in aircraft gas turbine technologies, with reduced weights lowering costs.

The first gas turbine power applications used paraffin-fuelled aero engines to drive relatively small electrical generators.

The prime movers and fuel were expensive, but they had a very small plant footprint, were quick to start, easy to use and clean to operate, ideal for a quick emergency power supply for example for larger power stations needing to restart.

Initially the turbines had low output ratings, 1–5 MW, and short lives, their thermodynamic performance was poor with efficiencies no more than 10%–20%.

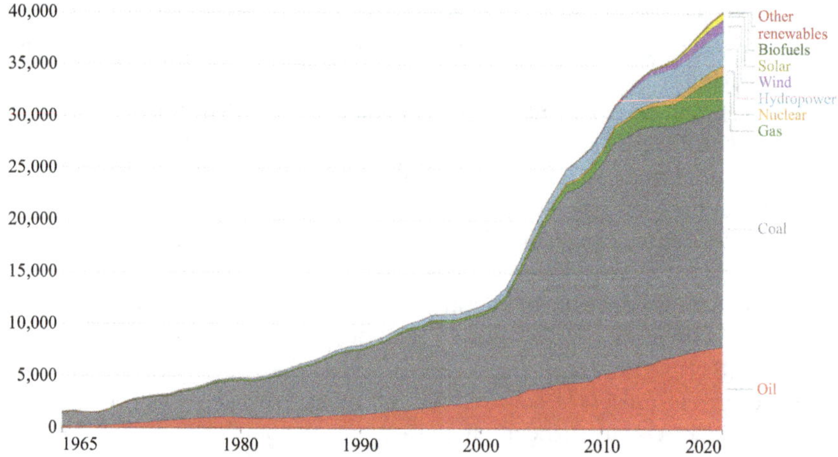

Figure 9.4 Energy consumption by source, 1965–2020, China

Figure 9.5 Datang Tuoketuo power station, China, currently the World's largest operational coal power plant. Output rating at full power of 6.7 GW, 2021.

However, the technology improved, stimulated by oil and gas industries needs for gas turbine electrical power sources of 5–50 MW, particularly offshore, with power input initially from aero gas turbines, supplemented by the larger power demands of the electrical generation industry.

The oil and gas industries found these gas turbine electric power sources, with outputs from 5 to 50 MW, invaluable for local power at well-head sites, particularly offshore, where there were weight restrictions.

Electrical grid operators then started to apply the same technology onshore initially for providing rapid power support from 5 to 75 MW, in the event of main grid failures and blackouts. Therefore, many thousands of relatively light-weight units, based on this technology, have been installed world-wide, gaining considerable experience of gas turbine manufacture generation.

Whilst these power generators were being installed, however, the gas turbine industry developed larger, more rugged, thermally efficient and compact industrial gas turbines, based upon steam turbine technologies. Learning from the marine and power industries, these machines had powers up to 150–200 MW and efficiencies up to 20%, and, in addition, they could be fired using cheap crude oils.

In the early 1980s, these larger industrial gas turbines appeared on the market, initially from GE but quickly followed by ABB, Mitsubishi and other manufacturers, with application for lower cost electrical power generation.

The thermodynamic imperative to raise power generation plant efficiency leads to the development of two-stage turbines, where the exhaust from the leading industrial gas turbine was used to raise steam in a fast flash boiler, powering a following, smaller steam turbine, see Figure 9.6, known as a Combined Cycle Gas Turbine (CCGT).

Each gas or steam turbine can drive a separate electrical generator, but in more sophisticated designs, both turbines can drive the same generator on a single shaft, lowering capital cost.

However, a single shaft solution, Figure 9.6(b), can increase operational risk and possible operating costs.

Many CCGT developments arose from lessons learnt in the oil and gas industries, with from the 1970s large offshore installations needing electricity generation using local fuel. This needed supply from light-weight power units of outputs from 1 to 100 MW mounted on offshore platforms.

Designs of combined cycle sets then developed rapidly, in USA, Europe and the Far East, for the supply of onshore combined cycle gas turbine power stations with unit outputs up 150–400 MW, fuelled from nearby gas sources. These were relatively light, compact machine arrangements and became particularly attractive for new fast-build power station developments.

Overall efficiencies of these new combined cycle plants are considerably higher than efficiencies of c 38%, currently achievable in large coal-fired steam plants, see Section 7.4.1.1.

The CCGT is a true heat engine, see Figure 9.6, the product of thermodynamic science, mechanical and electrical technologies, with a steam turbine, gas turbine and generator achieving overall efficiencies from fuel to electricity of up to 65%.

This arrangement and thermal performance would have delighted the scientist Sadi Carnot of eighteenth century, who laid our thermodynamic foundations.

It would also have charmed those nineteenth to twentieth century engineers, Watt, Stephenson and Parsons who pioneered reciprocating and turbine steam power and the scientist Faraday who pioneered the electrical machine.

CCGT power plant has also benefitted from the following twentieth century developments.

Use of an easily managed fuel, gas, which remained consistently cheap over a period of 20 years has yielded the following benefits:

- Increased thermodynamic efficiency;
- Reduced consequential fuel, coolant and waste costs;
- Reduced power station site footprints, with less construction, fast commissioning times and lower costs.

Although the above is now changing as a consequence of increased gas usage and the Russia/Ukraine war, the installed capital costs of these plants, ratings up to 0.8 GW, are much lower in £/MW than for large coal-fired or nuclear plants, with higher ratings, 1.5–3 GW.

However, the life cycles of combined cycle gas turbine power plants, to date, have been consistently lower than for large coal-fired or nuclear power plants.

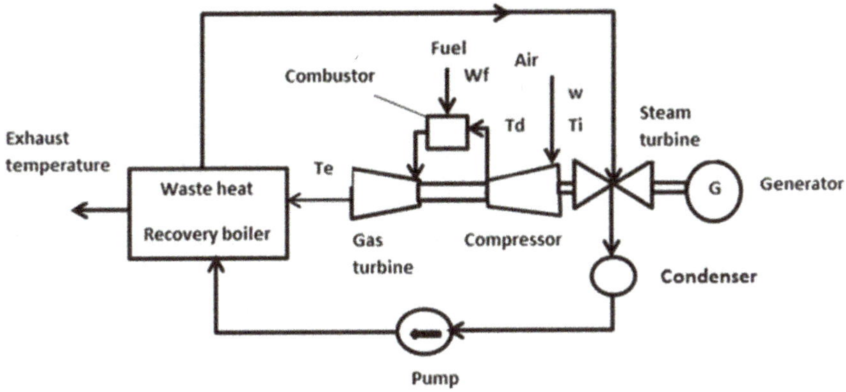

(a)

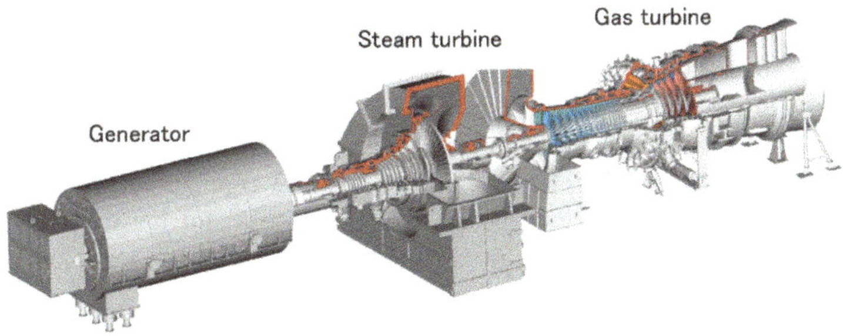

(b)

(c)

Figure 9.6 Compact Single shaft Combined Cycle Gas Turbine (CCGT) Power Station. (a) Diagram of a CCGT Power Station with single shaft, gas and steam turbines; (b) single shaft, gas and steam turbines and generator arrangement and (c) CCGT Power Station.

9.2.4 Waste to power and hot water

There have been many different schemes attempted to convert agricultural, industrial or human waste and crop waste economically into electrical and other forms of energy.

Difficulties in this sector arise from:

- Variability of the agricultural, industrial, human and crop waste and of the energy available from the waste stream;
- Need to concentrate and transport fuel material in batches large enough to make plant conversion economically viable, to maintain a reasonable conversion efficiency.

These issues can result in intermittency of the resultant generation operation.

Figure 9.7 shows a typical straw-burning energy recovery plant, although the raw fuel could also be wood-chip, domestic or animal waste.

The thermodynamic process at the heart of the plant has a relatively low thermal efficiency, 15%–27%. However, the plant makes good use of the waste and provides a local source of hot water.

The hot water can then be applied to industrial applications and large-scale food preparation or in cool regions would be useful for housing heating. So overall, the energy balance of the plant is improved.

It is interesting to note that many of these technologies were originally developed in Scandinavian countries and Russia, locations where ambient temperatures can be low, fossil fuels maybe scarce, but waste agricultural materials are plentiful. In those locations, heated water is valuable not only for agricultural and food preparation industries but also for housing.

The negative aspects of waste to energy schemes are as follows:

- Carbon release from non-biodegradable material, which could otherwise be stored in landfill;
- Air pollution as exhaust scrubbing is frequently too costly for small plant sizes;
- Residues from the incineration of ash and flue gas cleaning can leach into the environment.

9.2.5 Gas turbine or diesel engine generation plus hot water

Waste processes producing flammable gases that can be harnessed to generate electricity and extract heat, generally known as Combined Heat and Power (CHP) plants.

The prime mover for the smaller plants maybe a diesel engine but for larger gas supply streams it could be a gas turbine and this comparison is explained in Figure 9.8.

The advantage of these schemes, see Figure 9.9, is that they have a relatively low capital cost per MW, rely on a zero-cost fuel source but have a limited life span supplying electricity into the local electricity system. However, their environmental impact is low. In general, smaller schemes of 0.5–5 MW may involve a parallel array of small diesel prime movers, supplying low voltage electricity at say 400 V–3.3 kV.

(a)

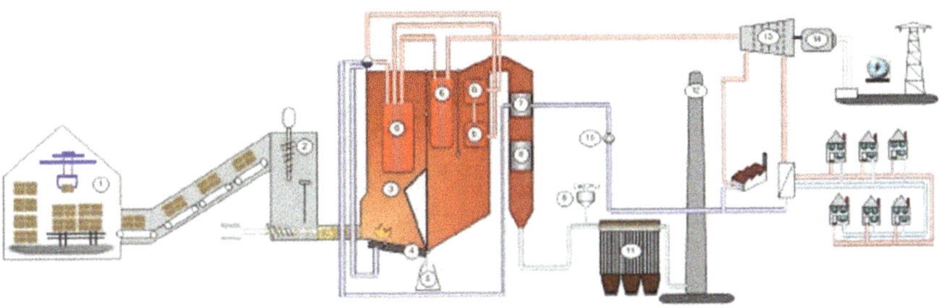

1 Straw storehouse
2 Bale-hoop-opener
3 Boiler
4 Vibration grate
5 Wet ash remover
6 Superheaters
7 Economizer
8 Air preheater
9 Dry sorption
10 Feed water pump
11 Fabric filter
12 Chimney
13 Steam turbine
14 Generator

(b)

Figure 9.7 Typical combined cycle gas-fired power station. (a) Compact 40 MW straw burning energy plant in Brigg, United Kingdom. (b) Typical schematic for a straw burning energy plant supplying electricity and heated water.

Large gas turbine schemes >5 MW may employ one or more parallel gas turbines, supplying HV electricity at up to 11 kV.

Figure 9.10 shows a modern offshore gas turbine driving a 53 MW electrical generator offshore.

GAS ENGINES vs TURBINES

	Gas engine	Turbines
Heat to power need ratio	1:1	2:1
Type of heat needed	Hot water, some steam	Mostly steam
Load	More variable	More constant
Electricity cost driver	X	
Heat/cooling cost driver		X

Figure 9.8 Prime movers for CHP schemes, their relative advantages

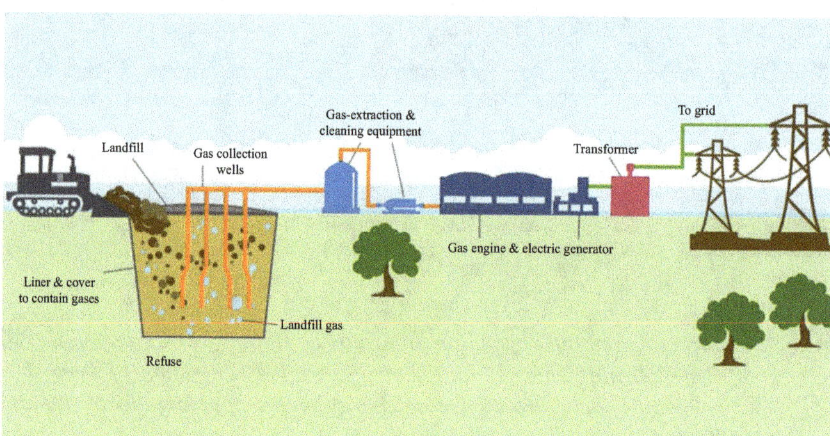

Figure 9.9 Landfill gas-powered diesel generation scheme

9.2.6 Carbon capture, storage and sequestration

This is the practice of extracting CO_2 gas from coal-fired power plant exhausts and capturing it by chemical processes or storage.

Figure 9.10 A GE LM6000 aero-derivative gas turbine driving an electrical generator offshore

Some Carbon Capture and Storage (CCS) projects have reused captured CO_2 by pumping it into reducing oil fields to pressurise them and extract remaining oil volumes, known as Enhanced Oil Recovery (EOR). In recent years, EOR has accounted for about 73% of global CO_2 capture, estimated at 28 Mega tonnes of CO_2 captured globally, out of 39 Mega tonnes produced.

This quantity has been reinjected into declining oil fields for EOR to push more oil out of the ground or to sequester the CO_2. Therefore, EOR eventually leads to more CO_2 emissions, directly due to increased hydrocarbon burning emissions from EOR but also indirectly from the fuel emissions used to compress and pump CO_2 deep into the ground.

A further challenge is finding suitably large CO_2 storage sites, where the gas may not merely be used to push out more oil but for sequestration. It appears that the sequestration of CO_2 into formerly exploited salt reservoirs can seal the CO_2 for long periods of time. However, not all underground chambers are well sealed, as in former salt mines, and stored CO_2 can leak back into the atmosphere. Therefore, sequestered CO_2 would need to be monitored for years, maybe centuries, to ensure that it does not return to the atmosphere. This raises the risk of liability being handed over to the public, years after private interests have extracted profit from the initial EOR enterprise.

The risk for us with CCS-technology is that it could be used to extend fossil fuel infrastructure life beyond a cut-off point for constraining atmospheric carbon below catastrophic levels.

9.2.7 Modern pumped storage hydroelectricity

Pumped storage power stations were originally developed by utilities to deliver electrical power for short periods at times of high demand.

In the United Kingdom, they were developed in the 1950–60s to speedily provide extra generation capacity at times of unusually high electrical load, for example during Queen Elizabeth's Coronation or at particularly large sporting events, such as the Cup Final.

They are generally hydro plants that can reverse the operation of their turbines. Such a power station can then use its electrical machines as motors, to pump water up to a high reservoir, storing potential energy. This is currently known as pumped hydro energy storage (PHES).

When required a PHES plant can then discharge that stored energy through their turbines with their electrical machine operation reversed to act as a generator. The control of this operation is from the National Grid of each country, and there are financial incentives for the PHES Operator so that they can recoup the energy costs of the pump-up; this is generally achieved by the operator receiving high payments for the energy released during those periods of high demand.

In Figure 9.11, two types of PHES arrangement are shown, in Figure 9.11(a) an open-loop pumped storage system, operating into a larger conventional hydro plant reservoir or in Figure 9.11(b), a closed-loop, standalone pumped storage station.

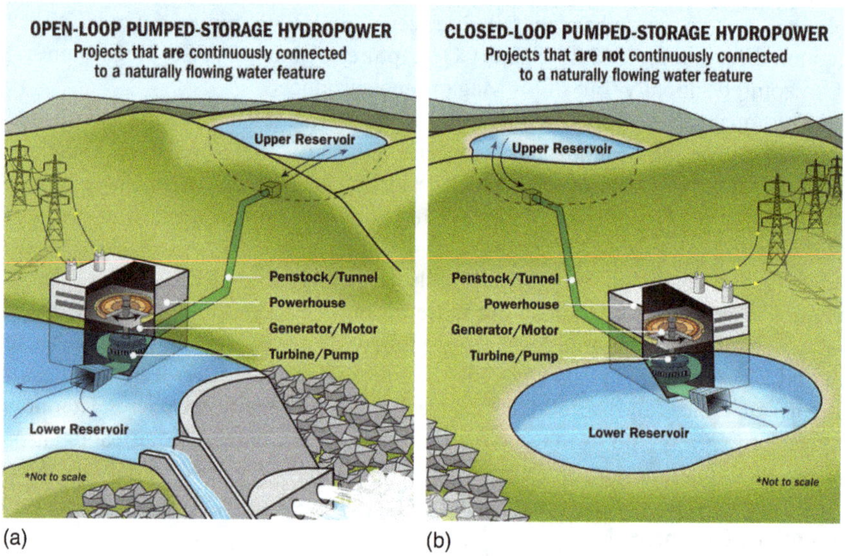

Figure 9.11 Examples of Pumped Storage Schemes. (a) Open loop pump-storage as part a conventional hydro-power plant and (b) self-contained closed-loop pump-storage scheme.

The advantage of an open-loop PHES is that its operation and maintenance costs can piggy-back off the output revenues of a much larger hydro plant installation, whose overall capital cost will be very large but can be defrayed over a long life.

The advantage of a self-contained, closed-loop PHES is that it involves a smaller initial capital outlay and can, in a mountainous location, provide a solution to local Grid power control issues.

Figure 9.12 shows the successful UK PHES at Dinorwig, Wales, completed in 1984, an example of a self-contained, closed-loop pump-storage scheme, originally

(a)

(b)

Figure 9.12 Self-contained, Hydro Pumped Storage Power Station Dinorwig, United Kingdom, rated at 1,800 MW with a storage capacity of 9.1 GWh. An example of Figure 9.10(b). (a) Upper reservoir, showing the intake area; (b) generator hall, showing the six generators and their exciter.

designed to stabilise national Grid AC frequency during sudden high electrical demand, due to social incidents. However, in recent years, it has been used on an almost daily basis to stabilise the Grid when handling high but variable levels of wind energy, particularly noticeable with the expansion of larger offshore wind resources.

At a PHES plant, turbine and generator rotation reverses at the changeover from generating to pumping or motoring, reversing the generator's electrical control and the water turbine hydraulic guide vanes control. Learnt experience has demonstrated that very rapid transitions from pumping to generating, and vice versa, can now be achieved by fast acting electronic control of the generator supply terminals and the hydraulic guide vane control, which has become the modus operandi of modern PHES plant.

Modern PHES plants have come to be seen as more reliable sources for stabilising grid AC frequency than innovative battery solutions, which so far have proven undersized, expensive and unreliable, although that is likely to change and improve with time.

European countries have been using PHES plants to stabilise AC grid frequency over different frequency zones, where intermittent or variable renewable electricity sources have become a significant part of European-wide electrical grid operations, alongside more fixed fuel resources, such as large coal and nuclear plants.

PHES plants of this type are therefore now widely used in Europe, India, China and USA.

Worldwide, there are currently 75 PHES plants installed, rated at >1,000 MW, with a further 50 planned or under construction, a significant proportion are in China, which has a large hydro resources and high variable industrial demand connected to the AC grid.

The rise in the application of PHES plants is driven by the rapid rise of variable solar, wind and other renewable sources of power, affecting operators' ability to control the fixed frequency of the AC grid.

PHES plant efficiency depends on:

- Turbine efficiency;
- Generator efficiency;
- Head loss in turbine intake and discharge channels, which in turn are dependent on their length, increasing with reservoir head height.

The round-trip energy efficiency of PHES hydroelectricity is reported as varying between 70% and 80%, with some operators claiming up to 87%.

The main disadvantages of PHES are the specialist nature of the site, needing both geographical height and water availability, and the high capital construction and plant costs, compared to more conventional diesel, steam and gas turbine plant. However, experience shows that PHES plants have a long life, for example in United Kingdom to date, the following plants have accumulated the operational experience listed in the following:

- Dinorwig 38 years;
- Foyers 53 years;
- Cruachan 57 years;
- Ffestiniog 59 years.

Therefore, PHES plants have lifetimes that can readily defray their higher capital cost and still earn long-term profitable income over many years. This reflects the long life of hydro-electric plants mentioned in Chapter 8.

9.2.8 Modern heat pumps

Air-Source Heat Pumps, ASHP, Figure 9.13(a), are used to move heat between two heat exchangers:

- One outside a building, fitted with fins through which air is forced using a fan;
- The other which either directly heats air inside the building, or heats water which is then circulated around the building through radiators or underfloor heating releasing heat into the building.

Such devices can also operate reversibly, in a cooling mode, where they extract heat via the internal heat exchanger and eject it, using the external heat exchanger, into the ambient air.

Some ASHPs can be used to heat water for washing, which is then stored in a domestic hot water tank.

Air-Source Heat Pumps are relatively easy and inexpensive to install and have therefore historically been the most widely used heat pump type.

The performance of Heat Pumps is expressed in terms of a Coefficient of Performance (COP). In mild weather an ASHP COP may be around 4, while at temperatures below −7 °C, an ASHP may still achieve a COP of 3.

While older ASHPs performed relatively poorly at low temperatures and were better suited for warm climates, newer models with variable-speed compressors remain highly efficient in freezing conditions allowing for wide adoption and cost savings in colder places.

Ground-Source Heat Pumps (GSHP), Figure 9.12(a), draw heat from the soil or groundwater, which remain at relatively constant temperature throughout the year, below a depth of about 9 m.

A well-maintained GSHP will typically have a COP of 4 at the beginning of the heating season and a seasonal COP of around 3 as heat is drawn from the ground.

GSHPs are more expensive to install than ASHPs, due to:

- The need for drilling boreholes for vertical placement of the GSHP;
- Piping and digging trenches needed for horizontal placement of piping carrying the heat-exchange fluid, usually water with some antifreeze.

A GSHP can also cool buildings during hot days, thereby transferring heat from the dwelling back into the soil via the ground loop.

(a)

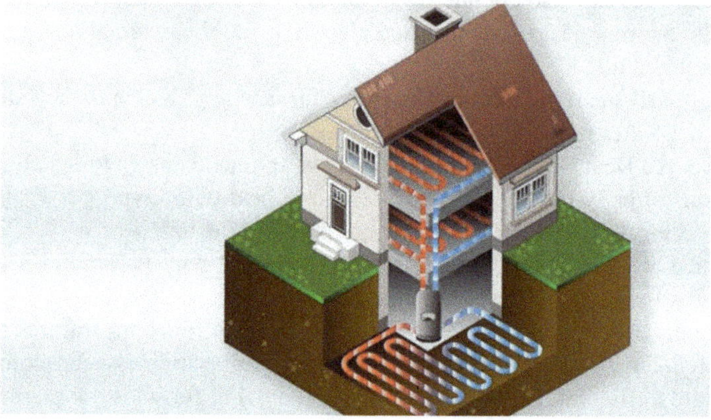

(b)

Figure 9.13 Heat pumps. (a) Air-source heat pump for a home; (b) ground-source heat pump for a home.

Solar thermal collectors or piping placed within the tarmac of a parking lot can also be used to replenish the heat underground to a GSHP.

9.2.9 Overview

This section has shown that existing energy technologies can contribute to a cleaner environment in the generation of electricity, through the significant efficiency improvements shown in Table 9.1.

Table 9.1 Conversion efficiencies, fuel to output, past and present with clean energy

	Plant type	**Output range (MW)**	**Fuel to power conversion efficiency (% when operating)**	**Operational cycle**
Early twentieth century	Railway steam locomotives	6.3–8.1 (US largest)	5	Continuous
	Maritime steam-ship propulsion	11.1 (TITANIC)	8	Continuous
	Coal-fired steam power station	10	10	Two-shifting
Today	Nuclear-fired Single Cycle	1,500–3,000	NA	Continuous
	Waste Plant with Hot Water Supply	10–50	15–27	Intermittent
	Diesel or Gas Turbine Plant Generation with Hot Water Supply	0.5–10	15–30	Continuous or Intermittent
	Coal-fired Single Cycle	1,500–3,000	30–38	Continuous
			27-35	Two-shifting
	Gas-fired Combined Cycle	500–1,000	50–65	Continuous
			45–55	Two-shifting
	Pumped hydro energy storage (PHES)	300–1,500	70–80	Intermittent, round-trip
	Air Source Heat Pumps	1–4 kW	COP 3–4	As required
	Ground Source Heat Pumps	1–40 kW	COP 3–4	As required

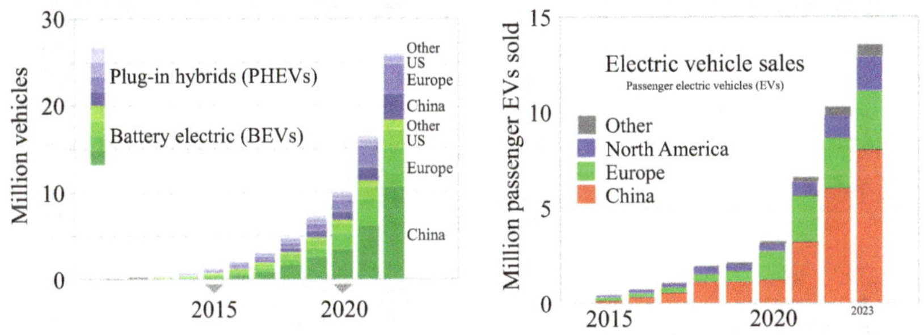

(a) (b)

Figure 9.14 Electric vehicles, numbers and geographical sales. (a) Numbers of electric vehicles operational and (b) numbers of electric vehicle sales.

9.3 Electric vehicles

The use of heat engines to drive cars, lorries and trucks has long been a source of concern, from the point of view of pollution.

Considerable progress has been made in the last 20 years in the development of both small and large electrically powered vehicles, designed to reduce the pollution caused by petrol- or diesel-powered vehicles.

The main challenge has been to provide sufficient electrical energy storage potential to give an electrically powered vehicle sufficient range for practical operation.

The development of lightweight but expensive lithium-ion batteries has provided a solution, which appears, at the moment, to be successful. The use of lithium-ion batteries has become more practicable as larger, lighter battery-packs have been developed and vehicle sales have steadily increased reducing battery-pack costs. However, lithium-ion battery packs have some limitations especially in terms of their large size and weight for larger, more powerful vehicles such as lorries and trucks.

Figure 9.15 Lithium-ion batteries and types of electric vehicle. (a) Lithium-ion battery; (b) price and market size for lithium-ion batteries; (c) large numbers of electric vehicles operational and (d) lorries and trucks.

Figure 9.14 demonstrates the increase in electric vehicles appearing.

Figure 9.15 demonstrates the lithium-ion battery, its cost per unit size from 1992 and types of electric vehicles which use this technology.

This is a growing industry, and some countries, notably China and Europe, are playing a very significant part in the expansion of these industries.

9.4 Summary

This chapter has demonstrated that modern generating processes, based upon earlier non-renewable and polluting technologies, but now using a variety of fuels, can and do attain better efficiencies and lower environmental impacts than traditional heat engine sources from the last 100 years.

At the end of the nineteenth century, heat engine efficiencies of 5%–10% were being achieved.

In the twenty-first century, efficiencies ranging from 15% to 80% are now being achieved, according to the heat engines used and the local temperature environment.

New air- and ground-sourced methods for extracting heat from the air or soil have become successful heating and cooling devices.

This demonstrates that engineering and thermodynamic improvements in human conversion and usage of energy can and are being made.

However, the volume of energy being used for human needs is now enormous, and human influence on the climate and our surrounding environments is now obvious.

Fossil fuel usage remains problematic from the point of view of global warming.

There is no doubt that efficiencies can rise higher, fossil fuel usage can change, and more improvements are in-train but the avoidance of fossil fuel usage is essential for clean energy solutions.

It is clear that the electrification of transport by road and rail will play a very large part in reducing CO_2 emissions from large global economies.

The next chapter will demonstrate what new renewable technologies for clean energy solutions are available and could possibly be expanded in the future.

Chapter 10

Renewable energy technology examples – present

10.1 Introduction

In his historical study of technologies, Cardwell [12] identified a distinction between:

- 'Concentrated' energy sources, such as nuclear or fossil fuels: coal, oil and gas;
- 'Dilute' energy sources, such as renewables: wind, tidal, wave and solar.

He drew attention to the problems of gathering and distributing energy from such dilute sources. Modern experience with renewables suggests that Cardwell's latter so-called dilute option is not only acceptable but also probably the most important lesson from renewables. We can and must develop our use of energy from 'dilute' sources, and there is no easy other way.

The carbon-dominated age, from 1700 to the present, extracted from concentrated energy in coal mines, oil and gas reservoirs, the use of roads, railways and pipelines, supplying factories, cities and homes.

However, from 1910 onwards, a growing electrical grid was developing to distribute power from concentrated energy sources, such as coal-fired and nuclear power stations and large hydro-electric schemes.

These electrical networks expanded in the late 1890s, initially in Europe and the USA, from relatively low voltages, e.g. 10 kV AC in the 1930s, up to voltages ranging from 100 to 500 kV AC.

More recently these networks expanded from the 1930s in the Soviet Union, South America, India, China and the Far East at voltages up to 1000 kV AC and including DC up to 500 kV.

These electrical networks are now enabled us to distribute energy from both concentrated and distributed fossil and renewable sources.

The value of these electrical networks has also proven practicable for incorporating 'dilute' renewable sources like wind, tidal, wave and solar, effectively exploiting the power electrical grid. However, that may have necessitated the construction of new sub-stations and a greater degree of on-line grid control.

Furthermore, it may be possible for other networks, for example wired or wireless, e.g. Wi-Fi and communications, to facilitate energy transfer from widely distributed, dilute, sources, to local customers.

This chapter will describe the development of predominantly 'dilute' renewable energy sources but also including some 'concentrated' sources. It brings into focus an aspect of renewable energy's Achilles heel, its lack of concentration.

Here I will show how this issue has been tackled in various renewable developments, the effect of 'concentrated' and 'dilute' sources the importance of storage. It will show the effect and costs of large-scale and local energy storage on renewables' development, so that we can have a clear view of what is possible for our energy future.

10.2 Solar

10.2.1 Overview

Humans have exploited solar power for the growing of crops over many millennia and in the last 150 years, and since the scientific revolution, have considered how the Sun's power could be harnessed in a more organised way.

The Sun's power is significant and reliable where humans live, particularly between latitudes 23 N and 23 S, it is subject to seasonal variation and weather effects but is a resource that can be, exploited and managed.

This section will demonstrate how that solar power knowledge has advanced for energy production.

10.2.2 Exploitation

10.2.2.1 Distributed thermal

In the twentieth century, small-scale solar power has been harnessed for homes, with the process accelerating in the 1970s, using technology for solar power to heat water for domestic use that could easily be mounted on flat-roofed buildings.

These small-scale devices provided valuable water heating capability in Southern Europe, particularly Greece and Spain, in North Africa, particularly Algeria and Egypt, and in China. These were locations where electricity was scarce and expensive but living and health standards were improving and there was a need to improve washing and bathing.

Figure 10.1 shows a simple solar power unit that can heat and store hot water on a flat roof and hot or warm water can flow down to a basin, shower or even a bath in the house below.

The author can remember these thermal solar installations appearing on better-off homes in Nigeria in the early 1970s, and they were widespread on small houses and flats around Beijing (北京), China in the late 1980s.

10.2.2.2 Large-scale thermal

In the 1970s–80s, large-scale Solar Power Towers (SPTs) were being developed in various countries, including in France, N Africa and China. The various prototypes used a field of flat or curved mirrors to concentrate sunlight on a tall tower, where a heat transfer system extracted energy from the concentrated sunlight. For example, water or molten salt was heated to raise steam and used to convert the energy.

Figure 10.1 Small scale solar water heater and storage tank for a single house

Figure 10.2(a) and (b) shows such a plant in France being used to concentrate solar energy, but this design was for research, using the heat collected to carry out high temperature experiments, rather than for industrial heating and to convert energy for distributed consumers.

Figure 10.2(c), on the other hand, shows two solar plants at Dunhuang China designed to concentrate solar energy on an STP to raise steam. The plant on the left of the picture was the pilot plant, while that on the right was to provide heat for a full-scale steam raising plant to drive a turbine powering a 100 MW electrical generator.

These SPTs proved to be high in capital cost and costly to maintain; however, the principle of collecting and concentrating solar energy had clearly been demonstrated, and the problem now is how to develop this technique in an industrially effective way.

10.2.2.3 Electricity

The photovoltaic, PV, cell can be used to convert solar energy into electricity and its history is as follows:

- 1888, Russian physicist Aleksandr Stoletov builds the first cell based on the outer photoelectric effect discovered by Heinrich Hertz in 1887;
- 1904, Julius Elster and Hans Friedrich Geitel devise the first practical photo-electric, PE, cell;
- 1905, Albert Einstein proposes the new quantum theory of light, in a landmark paper, explaining the photoelectric effect;
- 1941, Vadim Lashkaryov discovered p-n-junctions in Cu_2O and Ag2S protocells;

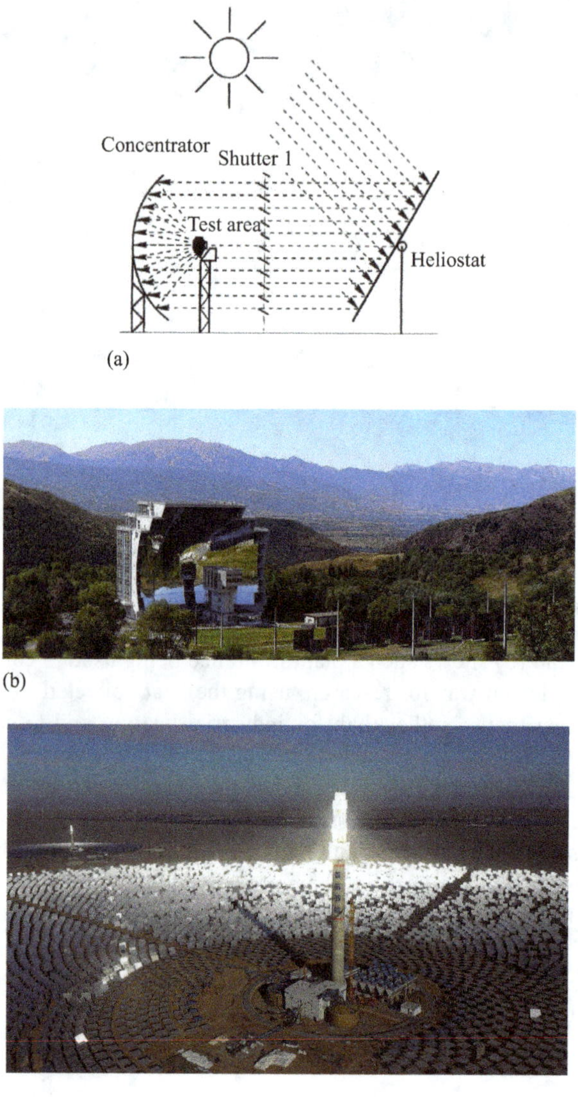

Figure 10.2 Solar energy collectors. (a) Solar furnace configuration, (b) solar furnace or SPT at Odeillo in France based on configuration in (a), (c) Dunhuang, China commercial-scale 100 MW SPT, showing large number of reflectors aiming solar energy at the tower.

- 1946, Russell Ohl patents the modern junction semiconductor solar cell, while working on advances that lead to the transistor;
- 1948, 'Introduction to the World of Semiconductors' states that Kurt Lehovec may have been the first to explain the PV effect in a peer-reviewed journal;

- 1954, First practical PV cell publicly demonstrated at Bell Labs;
- 1958, PV cells gain prominence by their incorporation into the Vanguard I satellite.

PV cells start to become available at affordable prices in the 1970s, arrays of PV cells, known as Concentrated Solar Power (CSP), can convert solar energy directly to DC electricity for subsequent conversion to MV AC, then transmitting MV AC power to a local substation for conversion to HV AC and connection to the main electric transmission Grid via substations, Figure 10.3(a).

Smaller scale PV arrays become usable for individual buildings, Figure 10.3(b).

(a)

(b)

Figure 10.3 Large- and small-scale solar energy collectors. (a) Spain's Solnova Solar Power Station. Foreground, three large CSP arrays, 50 MW each, with three substations, background, two SPTs, PS10 and PS20 cf Figure 10.1(c). (b) Smaller scale solar CSPs on a Hong Kong rooftop.

10.2.2.4 Developments

Small scale solar water heaters and storage tanks are still a valuable contributor in the developing world, and this manufacturing area has expanded enormously. The large-scale SPTs may prove to be white elephants, and the jury is still out. However, both large-scale PV solar farms and small-scale PV solar energy collectors on houses are growing in volume at a rapid rate.

10.3 Wind

10.3.1 Overview

The wind is changeable and sometimes violent, or even non-existent for significant periods of time; therefore, it is a complex resource to predict, exploit and manage.

In 1700 BCE, King Hammurabi of Babylon used wind-powered scoops to irrigate Mesopotamia. From 500 to 900 AD, the Persians used the wind to grind grain into flour. Others used the wind to transport armies and goods down rivers and the across seas and oceans. Sails revolutionised seafaring, no longer only requiring muscle power at the oars of galleys.

Therefore, because of its ubiquity, wind energy has been used by humans for more than 4,000 years, powering sailing craft, first on rivers, lakes and then the sea, harnessed on land to grind differing grains, pumping water and for cooling and ventilating buildings and habitations.

Many early human technical developments, in sails, ropes, lifts, mills and windlasses have advanced human wind power engineering skills.

More recently, mankind has used the power of the wind to pump water and produce electricity, so the idea of wind power, a natural source, is not new, and the wind has and will continue to play a prominent role in the development of human technology, being a known, if slightly capricious power source.

10.3.2 Exploitation

The modern, electrical exploitation of wind power started in the United States and Northern Europe in the late nineteenth century, in plains, rolling farmland and alongside rivers.

Recent developments were particularly energetic in Denmark, Germany and the United States. The principal initial motivation of these modern wind developments was to bring electricity to rural communities, where the land was relatively flat, and wind speeds, though variable, were significant.

Two famous pioneers were James Blyth and Poul la Cour, in Scotland and Denmark, respectively, developed the turbines shown in Figure 10.4, demonstrating early wind turbine engineering choices, should a turbine axis be vertical, the vertical axis wind turbine, VAWT, or horizontal axis wind turbine, HAWT, their perceived benefits being:

- Figure 10.4(a), VAWT, designed to have its vertical axis electrical generator at ground level, facilitating connection to batteries or local electrical supplies, and it was not necessary to veer or back the turbine;

(a)

(b)

Figure 10.4 *Early modern European Wind Turbines, for electricity production and battery charging. (a) 1887, Scotland, 10 kW, 7-blade VAWT, variable pitch turbine driving DC dynamo, developed by James Blyth. (b) 1891, Denmark, 10 kW, 4-blade HAWT, fixed pitch wind turbine driving a DC dynamo, developed by Poul la Cour.*

- Figure 10.4(b), HAWT, designed so that the horizontal axis electrical generator is driven at tower height. Power is then brought down the turbine tower, through flexible cables and the turbine must be veered or backed according to wind direction, in the same manner as traditional windmills.

10.3.3 Developments

There has been considerable discussion around the two types of turbines shown in Figure 10.1.

There are some advantages from the VAWT, because the generator is at ground level maintenance should be easier and the turbine does not need to be veered or backed.

However, most installed wind turbines now employ the 3-blade HAWT, shown in Figure 10.2, mostly with upwind blades, but there is still considerable discussion on the best way forward for the HAWT. Early experience on smaller wind turbines demonstrated that downwind turbines were easier to control and there are definite cost advantages in reducing the number of blades.

A study of Appendix 1.2: The Clean Energy Timeline shows clearly how wind power developed from the early work in United Kingdom and Denmark, up to the present day. The three centres for development growth in the wind energy industry now appear to be:

- Northern Europe;
- United States;
- India;
- China.

To date the 3-bladed upwind HAWT has proved the most widely used industrial design, developed and installed at onshore and offshore locations, with sizes ranging from 1 to 15 MW, Figure 10.2(a) and (b). Considerable prototype work was done in the United States and Germany to perfect large 2-blade downwind turbines, and the reasons for the modern 3-blade upwind HAWT choice for large turbines are probably rooted in the following:

- The greater proportion of development time having been devoted to 3-blade upwind HAWTs;
- Visual resistance to 2-bladed versions of HAWT turbines;
- Higher turbine axes lead to increased wind power production;
- Lack of operational experience with large down-wind HAWT turbines;
- Customer choice.

Recent wind industry power production experiences have shown that wind power has come of age and that turbine development is now of less significance than the manufacture, installation and operation of their large arrays.

For future wind power consideration will need to be given to the following:

- Lowering wind turbine costs;
- Speeding up and cheapening the process of wind farm installation;
- Developing better control and maintenance of large wind arrays;
- Improving integration of wind energy into the electrical grid.

The designs of larger wind turbines, 3–5 MW, have stabilised as operational experience has been gained, and larger turbines, 5–15 MW, are now being placed offshore where wind speeds are higher, more predictable and stable.

Many of these offshore turbines are 3-bladed with direct drive generators, Figure 10.5(b), and new installation lessons have been learnt. Much experience and personnel have transferred from the offshore oil and gas industries, but important lessons to cheapen those processes have been needed.

However, as wind turbine sizes and wind farm numbers increase, particularly further offshore where larger floating arrays and turbine ratings from 15 to 25 MW are now being considered, Figure 10.6. In that case concerns about weight, speed and location of assembly, simplicity of control and the irrelevance of visual impact, could lead to a reconsideration of present large wind turbine norms:

- Downwind, 2-blade HAWTs of lighter construction, Figure 10.2(c);
- Geared generator arrangements, lowering generator weight and cost;
- Both leading to reduced turbine and floating structure weight and cost.

An important part of wind power and renewable energy source development has been the increase in learning and knowledge in the area.

(a)

(b)

(c)

Figure 10.5 Large modern HAWT wind turbine deployments for electricity production. (a) 2014, Germany, BARD Off-shore Wind Project, 80 5MW 3-blade, upwind HAWTs, mounted on three leg steel jackets, for deep water. (b) 2026, United States, Coastal Virginia Offshore Wind Farm, 176 15MW Direct Drive upwind HAWTs. (c) 1954, Germany, a 2-blade downwind HAWT, developed by Ulrich Hutter, lightweight guyed tower and aerodynamic design.

A specific exponent of raising knowledge in wind power has been the Australian engineer, Rosemary Barnes, Figure 10.7 whose on-line channel 'Engineering with Rosie!' has been an inspiration for aspiring renewable energy buffs, including your author.

Figure 10.6 Floating offshore wind, different solutions, ranging from 15 to 25 MW output

Figure 10.7 Rosemary Barnes, a passionate exponent for renewable energies

10.3.4 Tidal

10.3.4.1 Overview

Tides are predictable and reliable but not necessarily conveniently accessible on every coast and island around the World.

However, in United Kingdom, we are blessed with being surrounded by the sea, with energetic tides.

Our understanding of those tidal effects has steadily improved over the last 300 years, based initially on human maritime experience from the fourteenth century.

The tides have been exploited as a power source for the last 200 years but that exploitation now needs to expand to embrace industrial requirements.

10.3.4.2 Exploitation

Current knowledge suggests several potential sites for large-scale tidal power installations, mainly in the Northern hemisphere, but with some in the Southern hemisphere, see Figure 10.8, [22].

Tidal flows in coastal estuaries in the Northern Hemisphere have been used since the seventeenth century to power mills for grinding corn and sawing wood. However, tidal technological learning has grown slowly as the resource is concentrated in those relatively few Northern Hemisphere coastal areas, with some isolated locations in the Southern Hemisphere.

In the second half of the twentieth century, tidal power was emphasised as a valuable renewable resource, because unlike wind power its contribution would be regular and predictable.

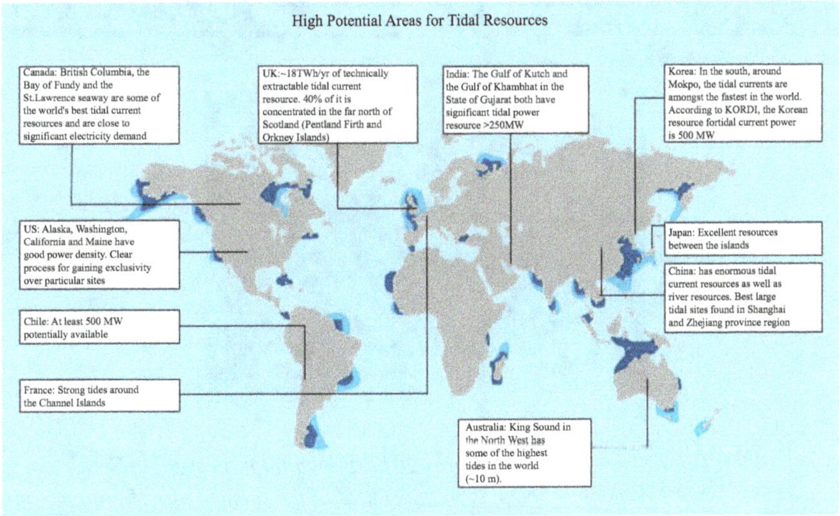

Figure 10.8 World Geographical Distribution of Tidal Resource Close to Shore, identifying promising locations [22]

There are three tidal power extraction structures:

- Fixed barrage structures;
- Tidal turbines, either on fixed or floating structures;
- Dynamic tidal power (DTP) structures.

10.3.4.3 Developments

So far, there have been three significant fixed barrage tidal schemes installed around the world, Figure 10.9, as follows:

- South Korea, Sihwa Lake Tidal Power Station.
  Opened in 2011, this is currently the world's largest tidal power installation, with a total power output capacity when completed in 2011, surpassing the La

Figure 10.9 Tidal Barrage Examples. (a) 2011, Sihwa tidal barrage in South Korea, 254 MW, powerhouse is in the centre of the picture. (b) 1966, La Rance tidal barrage in France, 240 MW, powerhouse is on the left of the picture. (c) 1968, Kislaya Guba tidal power station in Russia, 1.5 MW.

Rance Tidal Power Station. It is operated by the Korean Water Resources Corporation with a capacity of 254 MW.

The plant has 10 turbines each with a capacity of 25.4 MW and a capacity factor of 25%, with and annual electrical contribution of 552 GWh.

- France, La Rance Tidal Barrage.
  Opened in 1966, when it was the first and largest tidal power station in the world, operated by Électricité de France (EDF).

  The plant has 24 turbines of 10 MW a capacity factor with an installed capacity of 240 MW and capacity factor of approximately 24%.

  Operation of the plant has caused some progressive silting of the Rance ecosystem, which has affected some marine life in the estuary. However, EDF has made changes to the plant's turbine operation cycles, to reduce these effects and level out power delivery to the French Grid. This reduces the plant's peak output, delivering continuously an average 57 MW, with and annual energy contribution of 500 GWh.
- Russia, Kislaya Guba Tidal Power Station.
  A small tidal station located in Northern Russia close to Murmansk and the Norwegian border. The site was chosen because that coastline has several long, deep and narrow fjords from the sea, which could be dammed for a tidal project, the fjord walls effectively performing the role of the barrage.

  The station began operating in 1968 using a 400 kW French-built generator unit based on La Rance experience. This was later dismantled and the station closed for 10 years until funding resumed in 2004. Then a new 200 kW generator unit was installed, and in 2007, a second 1.5 MW generator unit was added. This station was designed as an experimental facility to demonstrate the feasibility of exploiting tidal movement for the generation of electricity around Russia's very extensive and indented coasts. This installation is a fraction of La Rance's and Siwha's power output, and there are no figures for an annual energy contribution, which could not be higher than 3.3 GWh.

Each of these three barrage schemes show what can be achieved at powerful tidal sites, demonstrating some of tidal energy extraction's benefits, but there are also some problems:

- Whilst tidal power is predictable in both location and time and is long-lived, it is confined to specific geographical locations;
- Tidal barrages require a large civil engineering commitment and high capital construction costs;
- Tidal barrage power outputs in MW are not large compared with other energetic power sources. Compare the power output of the La Rance and Sihwa, tidal power stations, c250 MW, with the power output of Drax, thermal power station, c3300 MW;

- Capital costs per MW are high; however, barrage and machinery lives can be long, as has been demonstrated in hydro-electric projects;
- As the tidal resource can be available over a long period, any installation needs a long life. Compare the lives to date of the tidal power station at La Rance (55 years) with the thermal power station at Drax (48 years).

There has been considerable ongoing interest in the above developments, particularly in the Northern hemisphere, in countries with an extensive seaboard needing reliable electric power sources, including United Kingdom, Canada, France, United States and Russia. But in every case the hard issue of the large upfront capital investment and the necessity for plant longevity has made investment difficult.

To mitigate capital costs, there has been a growing interest in tidal turbine devices, such as Figure 10.10. This trend has been driven, to some extent, by the engineering lessons of onshore and offshore wind but also based on tidal predictability. These tidal turbine schemes are seeking to avoid the heavy capital cost of

Figure 10.10 Tidal Turbine Examples. (a) 2008, Strangford Loch, SeaGen fixed and raisable twin tidal turbine, N Ireland, 1 MW. (b) 2017, MeyGen, Atlantis Resources fixed turbine, Scotland, 1.5 MW. (c) 2021, Fall of Warness, Orbital O_2 floating twin tidal turbine, Scotland, 2 MW.

barrage building by distributing a significant number of individual devices in proven, energetic tidal locations.

A great deal of this thinking has arisen from the deployment of wind power in United Kingdom, particularly offshore, where costs fell as carefully selected, planned small deployments were made initially in energetic but readily accessible offshore locations.

The tidal farm then becomes rather like a submerged wind farm, and important lessons, exemplified in Figure 10.10, have been learnt from these prototypes:

- Preliminary deployment of devices in small arrays increases learning, reduces risk and lowers capital costs, a direct result of onshore and offshore wind experience;
- Figure 10.10(a) and (b), turbines or devices that can be raised out of the water facilitate maintenance, reducing downtime and operational costs and reducing larger array costs;
- Figure 10.10(c), floating, moored structures have proved to be cheaper to construct onshore and are more easily deployed offshore, being easily detached for substantial repairs when needed. The orbital O_2 device in is an example, based on Scottish wave and tidal device development.

However, the most recent technology proposed for tidal energy is a half-way house between the barrage and stand-alone tidal turbines. It is known as the DTP dam.

None have been built to date, but the concept is based upon Dutch coastal engineering knowledge and experience.

A DTP consists of a T-shaped barrier, where the axis of the T is some km in length, built out from a tidal coastline.

This does not enclose an estuary but exploits coastal tidal action, without enclosing the sea in a costly barrage.

The main tidal movement along a coast runs parallel to it, and the mass of sea water is accelerated in one direction by the tide, against the DTP, then decelerated and reversed in direction.

The DTP dam is designed to be long enough to exert an influence on that coastal tidal movement, generating a water level differential, or head, between either sides of the dam. That head can be converted into power, using a series of conventional low-head hydro turbines embedded in the DTP.

The civil engineering costs of a DTP scheme, Figure 10.11, will be lower per MW than that of a fully enclosed tidal barrage, Figure 10.9, but higher than that of individual tidal turbines, Figure 10.10.

The costs of a DTP scheme will higher per MW than that of individual submerged tidal turbines; however, the long-term commercial benefits are likely to much better for a DTP than for distributed devices.

These three tidal technologies may all have a role to play in clean energy and their commercial costs and technical effectiveness need to be resolved by the tidal industry and energy community.

(a)

(b)

Figure 10.11 *Dynamic Tidal Power (DTP) structure. (a) Plan view of a DTP dam, showing the tidal head difference across the structure. (b) Tidal turbines embedded in DTP dam exploiting tidal head difference.*

10.3.5 Wave

10.3.5.1 Overview

Wave power is unpredictable and variable, but very energetic and usable on every coast and island around the world, see Figure 10.12 [22].

In United Kingdom, we are blessed by being surrounded by the sea with energetic waves, like our tides. Our interest in harnessing the power of the waves has developed during the past 50 years.

The first attempts to harness the power from the waves were initiated in United Kingdom by Prof Stephen Salter of Edinburgh University, who developed the prototype Duck device for extracting wave energy, Figure 10.13(a).

10.3.5.2 Exploitation

Current wave energy knowledge suggests many potential sites for large-scale wave power installations, around the World's continents.

However, realistically it is likely that the most valuable sites will be those closest to the wealthiest countries in the Northern and Southern hemispheres, Figure 10.12 shows that there could be potential around North America, Western Europe, China, Australia and New Zealand.

10.3.5.3 Development

The development of wave power is as yet less well-established than for wind or tidal power, because it has only recently been identified as a future potential source.

Figure 10.13 shows three wave energy generation schemes that have been developed, of which two have proven successful. However, there are still reasons for future optimism:

- Scotland, Salter's Duck, Figure 10.13(a).
  Developed in the 1970s and 1980s, this was a shaped device designed to respond automatically to wave motion and generate electricity. It was developed progressively in response to Government stimulation to be envisaged as a large, segmented array of moored but floating devices, located offshore in an energetic wave region, for example off the NW coast of Scotland and to be competitive against other large power sources, such as a nuclear power station.

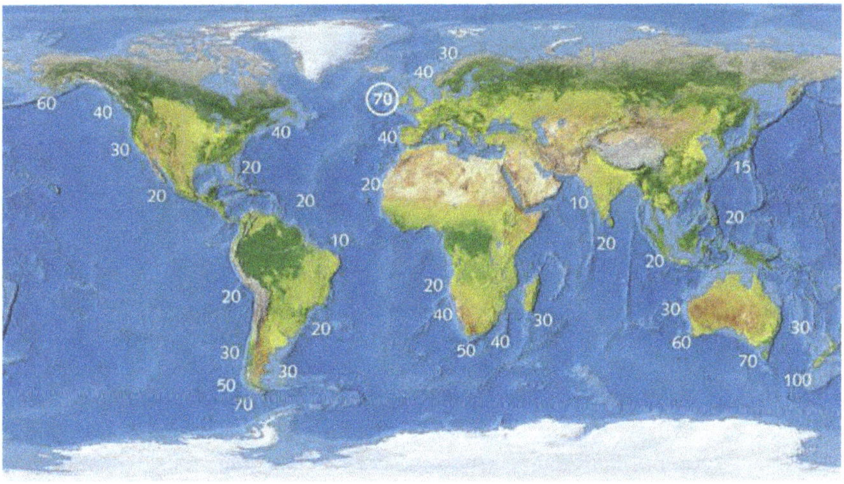

Figure 10.12 World Geographical Distribution of Wave Energy, figures in kW/m [22]

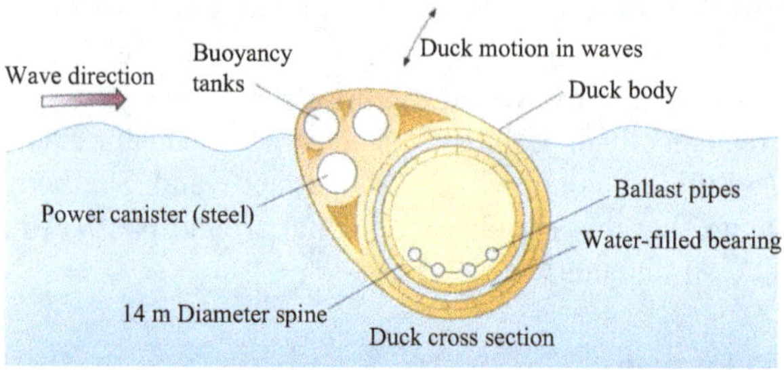

(a)

(b)

(c)

Figure 10.13 Wave energy examples. (a) 1976, Salter's Duck, Edinburgh University, aiming for 1 GW from a string of such Ducks. (b) 2009, Pelamis P2 wave snake, Scotland, 750 kW. (c) 2011, Wave energy production, WaveGen Wells turbines operational at Mutriku, Spain, 10 turbines installed and operating out of 16, full output 300 kW.

The engineering development of this technology, emanating for Edinburgh University, has had a large influence on renewable energy devices in the United Kingdom and Europe, because it stimulated so much analysis, many prototypes and many scientists, engineers and technicians with an interest in the subject. To date, the Salter's Duck technology has not generated energy from the waves but, in the light of other renewable solutions, this author suggests that wave power is still in the early stages of development.

- Scotland, Pelamis, wave snake, Figure 10.13(b).
  Several Pelamis devices of different powers and technology have been built, deployed and generated electricity to shore, either in a test environment or in early commercial offshore operation. Pelamis incorporated many lessons learnt from the Salter Duck programme.

  Three Pelamis P1 wave power devices were installed 5 km off the Agucadoura coast in Portugal in 2006 as part of a Portuguese consortium led by Enersis. The wave farm started delivering 2.25 MW of electricity from the three devices in September 2008, designed to provide electricity to 1,500 homes, using the state-run National Electricity Grid System.

  However, the three devices had to be towed back to the port after four months of commissioning because of technical problems.

  In 2008, financing of the generators' re-installation was made difficult for their customer in Portugal by the global financial situation, so the wave farm was shut down.

  A later development of Pelamis' wave devices, Pelamis P2, were operated by EON with a second device operated by Scottish Power Renewables. Both devices being installed at the European Marine Energy Centre (EMEC), Orkney, delivering 160 MWh to the EMEC grid before Pelamis went into administration;
- Spain, Basque Region, WaveGen Wells turbines, Figure 10.13(c).
  A small wave energy production site at Mutriku in Spain, an energetic wave area, uses 16 WaveGen Wells turbines to provide a local electrical power supply, based on a technology pioneered and developed on a single demonstration unit located at Islay in Scotland.

  The Mutriku site was chosen because a local inlet allowed a breakwater to be built, to house the devices and that inlet was close to local housing unconnected to an electrical grid.

  In the WaveGen device, up to 16 concrete turbine chambers contain unidirectional Voith-Siemens Wells turbines. These turbines rotate as waves push air into those chambers and continue rotating as air is sucked back out by the receding wave.

  Each turbine drives an electric generator supplying variable frequency and voltage AC, which then passes through a controlled power electronic converter to supply fixed frequency and voltage AC to local housing.

  Annual production from the Mutriku installation has been recorded as 2.4 GWh.

10.3.5.4 Weather-buoys

Wave power development is still in its infancy and, whilst there have been several failures, from the above sections, there have also been hard-learned lessons, and some successes, information from the Centre for Environment, Fisheries and Aquaculture Science (CEFAS).

Furthermore, wave power's history is still less than 40 years from the first prototypes.

Previous solar, tidal and wind experience all demonstrate that learning processes can be slow, and wave power's experience as yet is short compared to wind's 130 years or tidal's 56 years, with what is a very energetic but variable source, requiring extraordinary engineering ingenuity and resilience.

However, an interesting technical insight into the prospect for significant development of wave power resource is the experience of confidence elsewhere in the maritime field.

For example, developed countries, in Europe, North America, India and the Far East, have all devised valuable technologies for measuring and reporting wave height, wind speeds, magnitudes and directions around their coastlines. This development has generally been, of commercial necessity, for the better direction of large and small vessel ship movements in and out of their harbours, but also for the human exploitation of the oceans for leisure reasons.

Moored weather-buoys are used for this purpose, some are provided nationally, and some are installed by industries exploiting the marine environment, e.g. oil and gas companies.

Figure 10.14 shows a live image of the larger national and industrial weather buoys around the British Isles, showing swell directions and heights in metres, indicating the information available to mariners and operators of offshore equipment.

Weather-buoys providing marine data are powered either by solar devices and/or their own wave motion. They transmit their data at regular intervals either directly to shore or using UHF signalling to satellites and thence to shore. Weather-buoy data produced is typically:

- Temperature;
- Wave height;
- Wave direction;
- Wind speed;
- Wind direction.

This is generally available to mariners via the Internet, e.g. Figure 10.14(b), and the costs for operating the systems are recovered either directly from governments or by data charges to mariners for interrogating that data. Weather-buoys maybe less energetic than wave power technologies but involve some of the technologies and reliability needed for wave power. In terms of reliability, it is interesting to note the longevity of these weather-buoys, some of which have been in service in United Kingdom for 10–20 years, important information for those developing wave power systems.

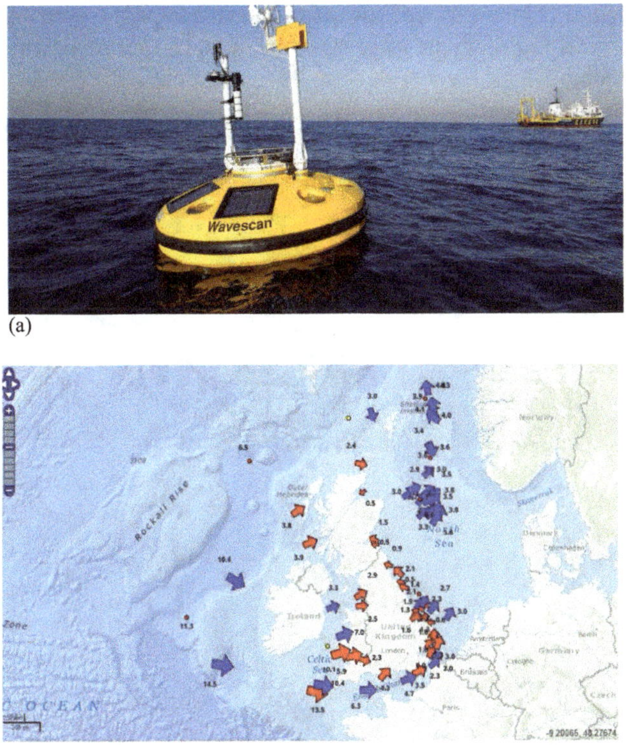

(a)

(b)

Figure 10.14 Wave-buoys as an exemplar for the development of offshore wave power. (a) Wavescan Wave-Buoy physical example; (b) UK Wave-Buoy data example, wave height and direction available on-line from CEFAS.

10.4 Hydropower

Hydro power stations have become a significant part of energy conversion since the 1880s, see Figure 5.12.

Modern hydro-electric schemes, in China, Africa and South America are now enormous plants, see Figure 3.10.

Many of these schemes were designed to control river floods, and electrical generation becomes a valuable side-product, paying for the dam and its controlling effect, exemplified by the huge Three Gorges Dam in Sichuan Province China, Figure 10.15.

The Three Gorges Dam was built between 1994 and 2003 across the Yangtze River (長江) made of concrete and steel it and its key details are as follows:

- 2,335 m long, 181 m high above a rock base and the dam top is 185 m above sea level;

- The structure contains 27 million m^3 of concrete for the dam wall, 463,000 tonnes of steel requiring the removal of about 102.6 million m^3 of earth during construction;

(a)

(b)

(c)

Figure 10.15 A modern hydro-electric power station in China, Three Gorges Dam, 2011. (a) San Xia Da Ba from the air. (b) Three Gorges Dam, San Xia Da Ba (三峡大坝) hydro-electric plant, China peak output 22,500 MW; (c) inside one of San Xia Da Ba's power houses showing six vertical hydro-electric AC generators, each rated at 750 MW.

- The dam reservoir is on average about 660 km in length and 1.12 km wide containing 393 km^3 of water with a total surface area of 1,045 km^2;
- On completion, the reservoir flooded a total area of 632 km^2 of land, compared to the 1,350 km^2 of reservoir created by the Itaipu Dam or the 5,250 km^2 reservoir created by the Aswan High Dam;
- When the water level is at its maximum of 175 m above sea level, 110 m higher than the river level downstream;
- Peak power output, 22,500 MW.

Large hydro-electric schemes are costly and involve huge national and international capital investments, but comparison between the Aswan High and Three Gorges Dams is interesting.

The Three Gorges Dam project was developed primarily to control excessive downstream flooding of the Yangtze River, the river's huge size means that the plant can also generate a great deal of electricity, 22.5 GW.

Whereas the Aswan High Dam was primarily designed to control the Nile for agricultural purposes. The modest electrical output of the dam, 2.2 GW, was an added bonus and valuable for the development of the nearby city of Aswan and the huge but distant capital of Cairo.

Hydro-electric developments are environmentally a contentious issue.

Environmentalists decry their disruption and resources their huge magnitude demands. There is no doubt that hydro-electric schemes can cause unforeseen environmental consequences. However, a side-effect of these complex financial and political preparations is that power station life expectations become extended for much longer than those achieved by fossil-fired installations. In fact, the operational steadiness of hydro-electricity assets has ensured their long lives, compared to fossil-fired plants, amortising the high capital costs.

So large hydro-electric plant lives can be extended beyond 60–120 years and more, higher figures are certainly hoped for at the Three Gorges and Aswan High dams.

The author's view is that hydroelectricity is a valuable tool in our reduction of the use of fossil fuels and that large hydro-electric plants will come to be seen as valuable energy assets. However, they are costly, must be well-planned and have significant influences on downstream communities.

10.5 Geothermal

Geothermal generation uses the thermal energy in rock formations experiencing ongoing volcanic activity. Energy is extracted from these rock formations by forcing water into the formation to heat it and raise steam to drive steam turbines.

Figure 10.16(a) shows a schematic of the principles of a geothermal plant and shows how such a plant can operate.

Figure 10.16(b) shows a plant in operation in Iceland using water injected into volcanically active rock formations to extract heat. This plant has an electrical

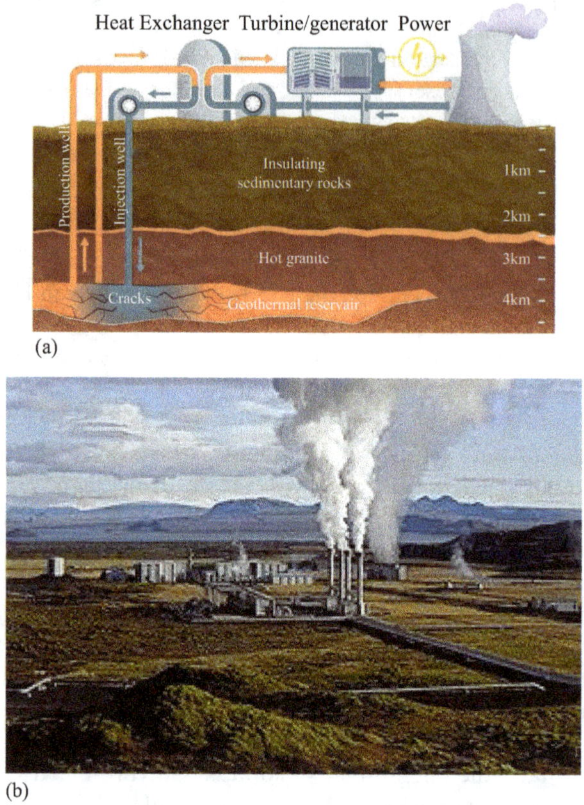

Figure 10.16 *Geothermal Power. (a) Schematic for a Geothermal Power Station. (b) Working example of Geothermal Power, steam rising from Nesjavellir power station in Iceland. Public domain: attributed to Gretar Ívarsson, Nesjavellir geologist.*

capacity of 120 MW and a thermal capacity of 300 MW for heating water for nearby housing.

Such plant processes are only possible in specific locations, but they are feasible in volcanically active countries, such as Iceland, Italy, Japan and New Zealand, where geothermal power is currently being exploited.

There are also engineering limitations as water forced into the rock formation can become contaminated with sulphur and that can be carried over into the heating systems causing corrosion, limiting life. However, these issues can be mitigated by feed water dosing. There may also be unexpected changes to the generation and heating plants' life duration due to unpredicted changes in volcanic activity with time.

10.6 Renewable enablement, small versus large

In 2019, world electricity generation by source was as follows:

- Total annual electrical generation 27 million MW-h;
- Using the following fuel sources:
  - 37% Coal;
  - 24% Natural gas;
  - 16% Hydro;
  - 10% Nuclear;
  - 5% Wind;
  - 3% Solar;
  - 5% Other.

One conclusion we can draw, in respect of renewable contributions, is that hydropower plants have taken an important place in our capability for generating clean energy, with the largest contributors being in South America and China, as follows.

Tidal power plants have perhaps been the next largest modern renewable power installations, with La Rance, France and Siwha, Korea being built between 1965 and 2004 with power outputs close to 250 MW.

However, despite tidal power renewable's effectiveness, its widespread application has eluded us. The principal reasons are, that fixed tidal plant can be very large but of long life, requiring huge capital investment at specific locations in particular countries. Such locations are not necessarily widespread, around the World either or in specific countries. Therefore, tidal power has not, so far, been a universal renewable energy production panacea. It involves large long-term investment in specific locations and commitment from central governments. In this respect, it has some similarities with hydro power.

On the other hand, wind turbine development has meant that relatively low power output turbines can be deployed, using repeatable units, in small or large farms, onshore or offshore, having relatively short lives, say 20–30 years.

Table 10.1 Summary of very large hydropower plants, worldwide

Hydropower plants	Country	Capacity (MW)
Three Gorges or SanXiaDaBa Dam	China	22,500
Baihetan Dam	China	16,000
Itaipu Dam	Brazil Paraguay	14,000
XiLuoDu Dam	China	13,860
Belo Monte Dam	Brazil	11,233
Guri Dam	Venezuela	10,235
WuDongDe Dam	China	10,200
Aswan High Dam	Egypt	2,200

Therefore, wind power has proved investable because it is relatively easy to develop, repeatable, with falling costs and limited life.

The development of offshore wind has given intense impetus to increasing the size of wind industry renewable installations and has been affected by many lessons from the offshore oil and gas industries, learnt in:

- The United States, in the Gulf of Mexico;
- Europe, in the North Sea;
- China in the Northern, Eastern and Southern Seas, Bo Hai, Dong Hai and Nan Hai.

The results of these lessons demonstrate that we can develop large-scale offshore wind farms of up to 1,500–2,500 MW, with intricate sets of cable interconnections, both between farms and for shore connection. These interconnectors are currently largely AC up to 350 kV but with increasing use of DC for wind farm to shore connectors of up to 300 kV.

Similar lessons to wind have been learnt with solar power. Solar power demonstrated its capability for development of large solar arrays, in the USA, North Africa and China, with outputs up to 1,000 MW. However, the capital costs of the interconnectors to distant electrical loads from these installations have been high, compared to other cheaper renewable installations located closer to energy load centres.

Therefore, early renewable wind and solar power in the late 1990 and early 2000s was generally of smaller size, with arrays ranging from 5 to 20 MW, but with technology that was expandable and investable.

The early production of small solar arrays, initially in Germany and Japan, brought down solar power unit costs. Then solar panels were developed in China, with a large home market, bringing costs down further. This lead to their widespread installation in many countries, with increased application, not only in developed countries, but also in poorer communities close to the Equator, where solar power enabled access to communications, through TV, telephony and mobile phones. This technology has been ground-breaking because it bypassed the costly electrical grid, delivering information and knowledge direct to communities, which are some of the poorest in the World.

So, renewable devices are not concerned only with their technology but also involve geography and application.

10.7 Storage for variable renewable energy

Storage has been an important issue for renewable energy because of the variable nature of many renewable resources.

Essentially, smoothing energy production has become an increasingly important issue for the development of renewable energy devices.

That may be smoothing of the rapid turbulent changes of air passing through a wind turbine or of the water passing through a tidal turbine, but it also includes the slow balancing of the gross variations in daily wind turbine power, input into the electrical grid.

Renewable energy engineers have therefore been particularly concerned to mitigate disturbances by the application of storage technologies as follows, see Figure 10.17. The main contenders for providing energy storage are shown in Figure 10.17.

Their contribution to supply is described below.

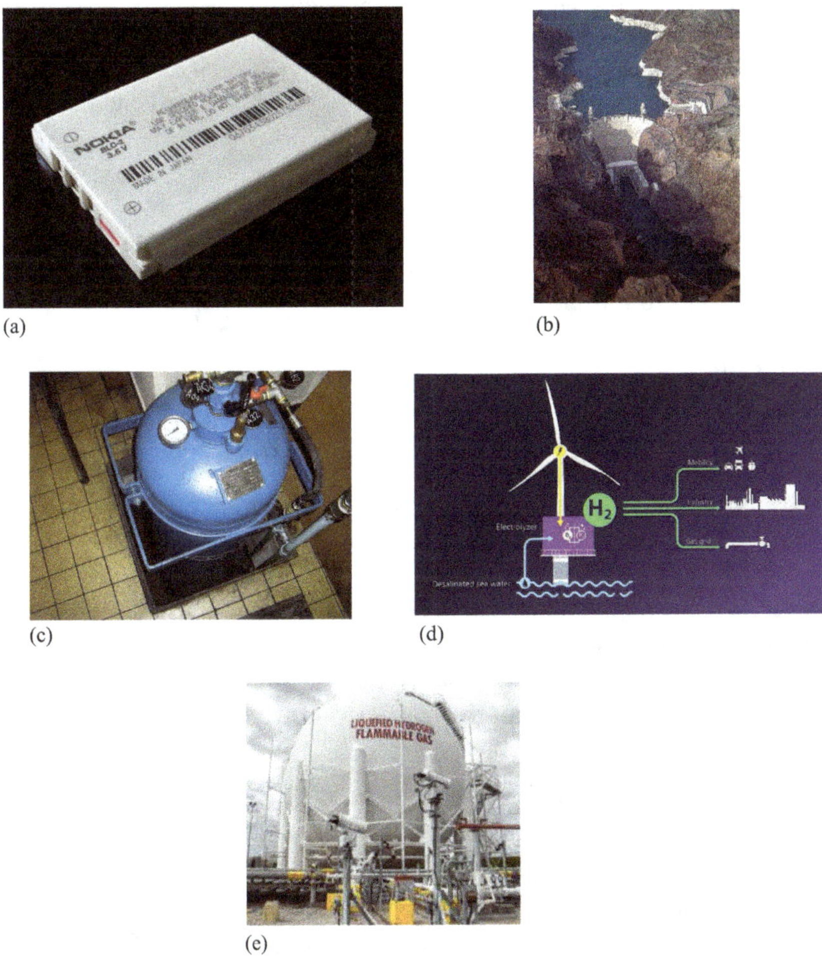

Figure 10.17 *Energy storage mechanisms from renewables using batteries, water, air or hydrogen. (a) Electric charge storage, Lithium-Ion Battery Pack; (b) water storage, Hoover Dam, USA; (c) air storage cylinder and compressor; (d) a novel proposal from Siemens Energy separating and storing H_2 using a wind turbine; (e) liquid hydrogen Tank, NASA, Kennedy Space Center.*

10.7.1 Batteries

Both large and small batteries have been proposed.

In the early expansion of wind power in the 1980s–90s, there was much talk of large-scale battery installations to smooth the variable power outputs of wind turbines, Figure 10.17(a).

In fact, most of these experimental efforts have had a negligible influence on the expansion of wind power.

However, there has been some expansion of hydro schemes to balance variable wind power, the most notable in United Kingdom being pumped storage schemes, Figure 9.10. Therefore, renewables variability has largely been accommodated not by batteries, but by the more flexible operation of hydro schemes and the operational controls of extensive AC electrical grids.

10.7.2 Water

The advantage of water as a storage mechanism is that it is dense, readily available, and relatively easy to manage, from our eighteenth century knowledge of hydrodynamics, Figure 10.17(b).

10.7.3 Air

Air is also readily available, but not dense, although it can be compressed, and relatively easy to manage, from early knowledge of aerodynamics, Figure 10.17(c). There has been considerable interest in harnessing and storing energy from the wind by using wind turbines to pressurise compressed air units.

10.7.4 Hydrogen

More recent technology has been proposed to use the power from a wind turbine to compress air and separate hydrogen for commercial use, Figure 10.17(e).

10.8 Renewables instrumentation and control

The control and instrumentation of distributed renewable resources embedded within larger energy transfer systems is an important area for current renewable energy research.

The top-down control methods developed in large-scale electrical transmission networks are inappropriate for these new systems now.

Greater use is now being made of the use of the internet, big data to deal with the distributed size of these assets as well bottom-up data collection methods and analysis techniques.

10.9 Summary

This chapter has set out the way in which renewable technologies have developed over the last century.

Some technologies, for example wind and hydro power, have reached back into earlier engineering development, adapting those ideas of power but perfecting them with modern materials and knowledge.

However, some of these technologies have been developed from recent knowledge, for example the application of semiconductors to solar panels, these being areas of completely new technology.

If we reflect on the development of fossil-fired technologies, in earlier chapters, before renewables readers will realise that there were many false steps.

The steam locomotive went through many evolutions before the optimum arrangement was achieved and even then there were variations between the products of different countries, due markets and fuel sources.

The internal combustion engine for cars and lorries, introduced at the end of the nineteenth century, went through many evolutions before we arrived in the twenty-first century with modern, efficient, high-performance petrol and diesel engines.

We are currently going through an evolution to develop modern electric motors for cars and lorries. Some of that technology is evolving from recent rail electrical power. But much more innovative ideas are being developed, in respect of permanent magnet materials, modern conducting wires and most especially for electrical control, batteries and charging are also necessary.

Chapter 11
Clean energy technology development – future

11.1 Introduction

Since the 1970s, climate change has become a social, political and industrial issue with new power technologies are being developed to resolve the problems that the human effects of climate change are causing, primarily the excessive production of CO_2 into the atmosphere due to the burning of the fossil fuels, coal, oil and gas.

These new power technologies may not necessarily be simple refinements of earlier technologies, such as those described in Chapter 9, but genuinely new conversion methods, supplying energy into future pathways for renewable human development needs, without relying on fossil fuels or creating consequential pollution, such as those described in Chapter 10.

These new technologies include wind, wave, tidal, solar and other, newer and perhaps more disruptive technologies, such as batteries and the integration of power and information networks. This chapter will demonstrate the progress made in these technologies and their potential for future success; however, the chapter starts by demonstrating how uncertain the introduction of new technologies can be.

There may be no certain outcome from these technological developments, which may be constrained by costs, material availability, effectiveness or simply social or environmental unacceptability. However, from some of these technologies, we must surely find a route to a cleaner energy future.

11.2 Technology developments-historical perspective

11.2.1 Technological changes in energy usage

Caution is needed when assessing new renewables technologies, because in some locations, they are in very early stages of development and deployment, so many lessons need to be learnt and are being learned.

That means that not all developments may be successful, and many improvements will need to be made before we can use these sources reliably for a long-term future. Therefore, not all renewable energy innovations entering the market may yet be serviceable or profitable. There may be significant failures in technology or capital facing us in this area in the future but there are valuable lessons to be learnt from earlier technological developments.

However, to be positive, I want to emphasise successful engineering endeavour by citing two specific historic examples:

- Steam locomotion, in the early nineteenth century, when various methods were proposed for the upcoming railway revolution, Figure 9.1;
- Aviation, at the start of the twentieth century, when many different types of aircraft were being proposed, Figure 9.2.

From the following commentary, and the sequential historical images, I want to demonstrate the extraordinary changes that developed from initial, and sometimes unpromising, technical developments. I also want to emphasise that such developments take time to achieve full operability and acceptance.

In the following two cases, for steam locomotion and aviation, it took 120 and 97 years, respectively.

11.2.2 Historical reviews

11.2.2.1 Steam locomotion

Steam-powered locomotion of wheeled road or rail vehicles started towards the end of the eighteenth century. The earliest development was in France, where high quality, surfaced National roads existed, encouraging engineers to develop steam driven road locomotives, see Figure 6.1(a).

In the United Kingdom, the developments came primarily out of the mining industry, where wooden and iron tracks had been developed for the haulage of coal and minerals. This was rather later than the French steam road locomotive.

The steam railway locomotive developed in the early nineteenth century as shown in Figure 11.1(b)–(e).

The 'Traveller', a steam-propelled walking locomotive of 1813, was an early innovation that failed in railway service. Although it did incorporate some valuable future lessons, particularly the two horizontal cylinders on either side of a horizontal boiler and reciprocation, it did not, however, incorporate Watt's ingenious mechanism for converting reciprocation into rotary motion, Figure 4.2(c) and (d), so its walking mechanism, whilst ingenious has not been repeated.

To stimulate railway locomotion in the United Kingdom in 1829, a famous trial was held on a track at Rainhill of the new Liverpool to Manchester Railway. The purpose of the trial was to select the best steam railway locomotive.

One prototype steam locomotive and its design stood out from all the others. That was the 'Rocket', designed and manufactured by George and Robert Stephenson, see Figure 11.1(e). It had many innovations, a horizontal multi-tube boiler, two double-acting cylinders with associated valve gear, and reciprocating gear to drive the wheels. Compared to its competitors it performed with flying colours, Figure 11.1(c) and (d).

Its nearest challenger 'Perseverance' failed to meet trial conditions on the day but did receive a consolation prize.

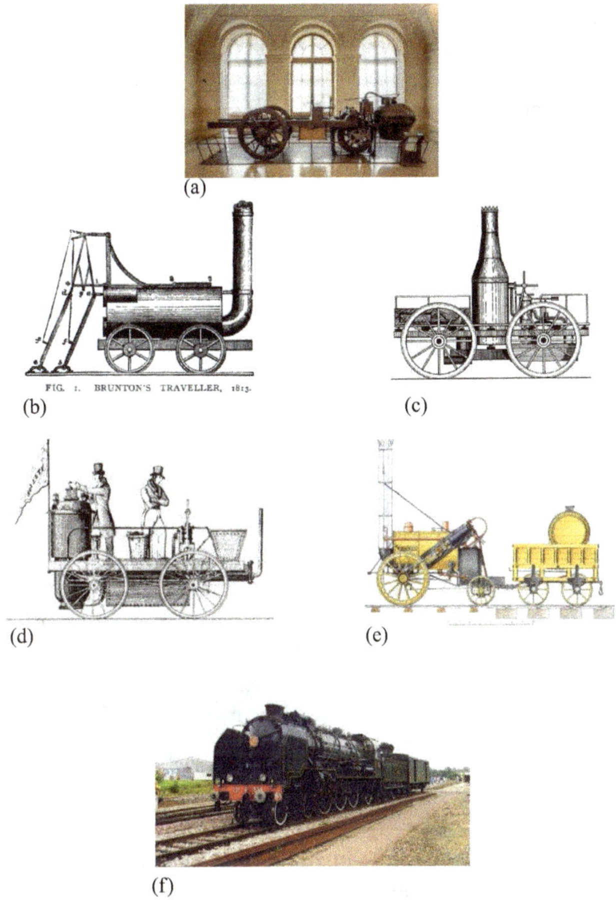

Figure 11.1 *Steam locomotion summary. Unsuccessful prototypes leading up to the success of Stephensons' 'Rocket' and world-wide improvement. (a) 1770, Cugnot's 'Fardier à Vapeur' 3-wheel road locomotive, vertical boiler and cylinders powering a driving wheel; (b) 1813, Brunton's 'Traveller', horizontal boiler and cylinders driving a walking mechanism; (c) 1829, Burstall's 'Perseverance', vertical boiler and cylinder, rolling element bearings. Runner-up at 1829 Rainhill Trials; (d) 1829, Ericsson and Braithwaites's 'Novelty', World's first tank locomotive, vertical boiler and cylinder; (e) 1829, Stephensons' 'Rocket', horizontal boiler and angled cylinders. Winner 1829 Rainhill Trials; and (f) 1960s, modern high speed French steam locomotive, Pinnacle of 120 years of thermodynamic and locomotive performance improvement.*

The lightweight and elegant 'Novelty' was the last locomotive to drop out of the trial; however, because it did not tow a tender for coal and water, it was an interesting design that subsequently evolved into the valuable steam tank engine.

The 'Rocket' went onto pioneer all subsequent fundamental steam locomotive design for the next 100 years. The pinnacle of that development is exemplified by the beautiful, powerful and thermodynamically efficient French steam locomotives, such as shown in Figure 11.1(f), still operating in the 1960s and hauling Le Train Bleu.

11.2.2.2 Aviation

Interest in aviation started in the early nineteenth century and successful ideas began appearing before the start of the twentieth century, Figure 11.2, key innovators being Cayley, Lilienthal, Blériot, Givaudan and the Wright brothers.

Cayley built model gliders and in 1843 proposed a larger flying aerial carriage, whilst Lilienthal built a functioning glider and Blériot developed a variety of designs. The first true flight was achieved by the Wright brothers in 1903, see Figure 11.2(e).

By the start of the First World War in 1914, several flying machines were available to all the combatants, some of doubtful safety and performance. But by the start of the Second World War in 1939, a very large range of sophisticated aircrafts were available in Europe, America and the Far East, see for example the Supermarine Spitfire, Figure 11.2(f), emphasising the sophistication of modern aeronautics at that time, echoed by many other machines of the time.

11.3 Problems and solutions

For those who are passionate about climate change the above lessons, from steam locomotion and aviation development, of slow progress with isolated success against many failures, maybe a bitter pill.

But they are an essential part of all practical development of all energy technologies.

This has been exemplified in renewables where, whilst photo-voltaics and wind power have progressed relatively rapidly, early work on tidal and wave power development has been slow and tough. Many devices have been proposed and prototyped, but there have been many failures and working prototypes disconnected due to the lack of ongoing funding.

The way forward for the wave and tidal industry is not yet clear, and these are lessons which any engineer could advise renewable developers to prepare for.

However, there have been successes, and we can learn and profit from them, as our forerunners did in the nineteenth century.

Just consider how long it took for the steam engine to develop from an expensive toy to a powerful efficient machine.

The following sections describe the current progress made by renewable energy, compared against existing energy sources.

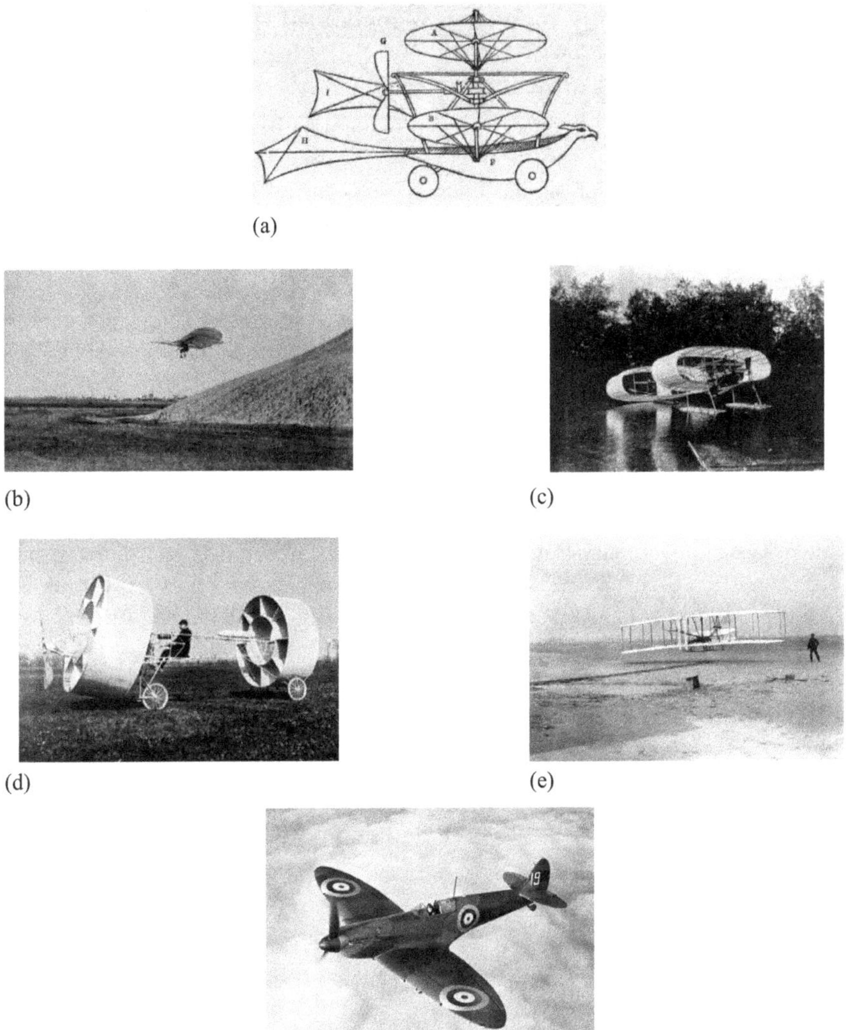

Figure 11.2 *Aviation summary, unsuccessful prototypes, Wrights' Flyer and wartime experience, modern improvements. (a) 1843, Cayley's proposed Aerial Carriage; (b) 1896, Lilienthal's successful glider flight; (c) 1906, Blériot's unsuccessful third prototype; (d) 1905, Givaudan's unsuccessful circular wing design; (e) 1903, Wright Brothers' Flyer. Ready for successful flight, Kitty Hawk, North Carolina, USA; and (f) 1940, Supermarine Spitfire fighter Mk1 in service in the early part of WW2, Pinnacle of 100 years of aviation performance improvement but superceded by the jet aircraft.*

11.4 Current energy technology-costs and trends

11.4.1 Energy from differing sources

The energies available from different fuels are summarised in Figure 11.3 in terms of energy quantity against specific fuel mass. Traditional fossil fuels such as coal, oil and gas produce rather similar energy per industrial fuel mass unit, whereas nuclear fuel produces a similar amount of energy from a much smaller fuel mass. On the other hand, renewable energy sources need to access very large masses of air or water to supply significant energy.

However, Figure 11.3 clearly shows that with current twenty-first century development, the outputs from renewable energy sources can be impressive. This, of course, is the result of huge recent development in those industries.

Patterson [27] in his book 'On Energy Electricity vs Fire: The Fight For Our Future' attempts to address this issue.

11.4.2 Capital costs, fuel and generation

Table 6.1 gives a summary from the United States of generation fuel source capital costs for 2019–20. It shows that gas, onshore wind and PV costs are low, and hydro-electricity, coal and PV are median, whilst offshore wind and nuclear costs are high.

Figure 11.4 gives a graphic of current capital expenditure (CAPEX) trends for N American generation plants in $/MWh from 2010 to 2019 for different fuel sources, from IRENA 2020.

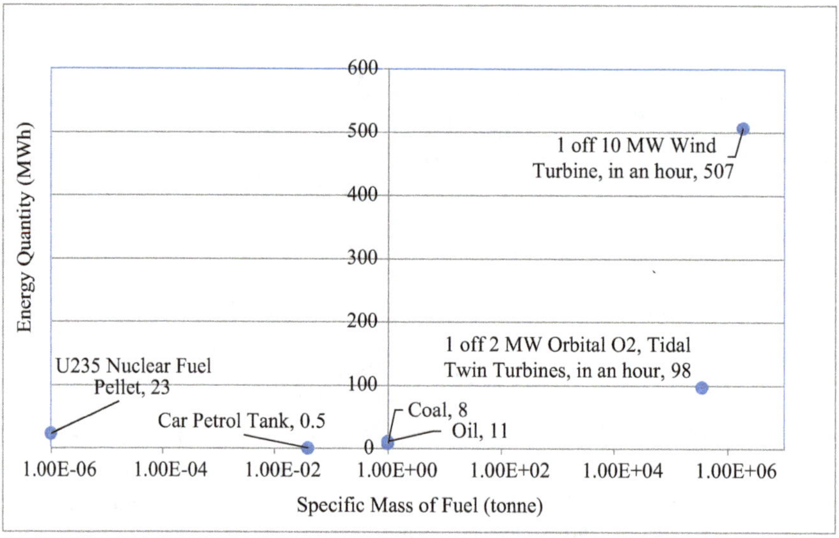

Figure 11.3 Energy quantity against mass in MWh/tonne, fuels from different sources

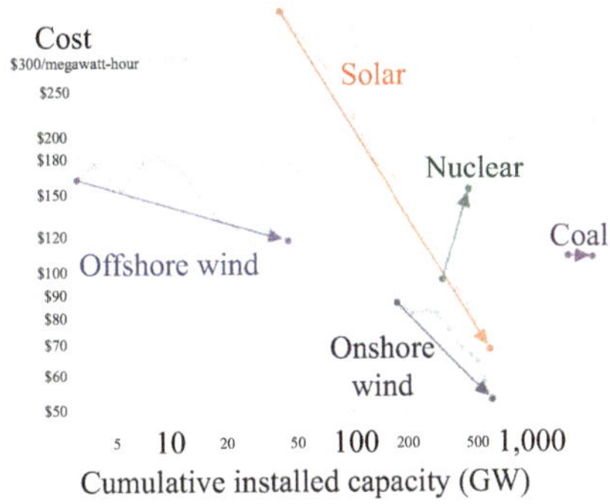

Figure 11.4 N America IRENA, CAPEX Cost of Energy trends in $/MWh, coal, nuclear, solar and wind, 2010–19

The following interesting features arise:
Coal-fired energy:

- Capital cost remains relatively constant over the period, because this is a mature fuel and few improvements have recently been made;
- Nuclear power:
  Capital costs are rising, primarily because the nuclear supply chain has not been stable and the regulatory consequences from earlier accidents, Three Mile Island, Chernobyl and Fukushima. National and world nuclear regulatory authorities have gradually raised the necessary safety requirements, increasing capital costs. This also increases operational costs for the large fleet of installed nuclear reactors around the world due to additional operational legislation. However, there has also been an increasing standardisation of plant designs, concentrating on reliable pressurised water reactors, originally developed to power nuclear submarines. As a consequence of these costs, nuclear projects are, at present, relatively few and far between, although China has a large and growing programme of construction;
- Renewables, wind and photo-voltaic solar:
  Costs are falling fast as increasing capacity reduces manufacturing, installation, capital and operational costs and few safety or design issues have retarded these developments.

11.4.3 Levelised cost of energy

Table 6.2 gives a tabular analysis of the Levelised Cost of Energy (LCOE) of energy production in the European Union in €/MWh, over a range of fuel sources from fossil through renewables to nuclear power, with upper and lower costs from 2009 to 2021, demonstrating several interesting features, as follows:

- Coal and gas:
  In earlier years, these fuels had low LCOEs. However, LCOE has risen recently. This is probably due to a combination of less energy being extracted,

lower investments being made in the fossil-fired industry and perhaps because of stronger regulatory control;

- Renewables, Wind and Photo-Voltaic Solar:
  In their early years, these assets had high LCOEs. However, as new technologies emerged, costs have fallen. This is due to the mass manufacturability of PV panels and wind turbines, where capital costs have reduced as larger assets are installed in increasing numbers. It also suggests that their reliability has risen, as installed asset numbers increase, due to rising component reliability and, with experience, falling maintenance costs;
- Hydro-electricity:
  Only 1 year's data is shown, but one can draw some conclusions.
  Hydro-electricity assets have a very high capital cost, due to a dam's huge civil engineering structure. That elevates LCOE in the early years of plant operation.
  Smaller hydro-electricity plants have higher capital costs per MW than larger plants.
  However, hydro-electricity assets can have very long lives, from 60 to 120 years, so capital costs become amortised over a longer period than other energy assets and maintenance costs do fall. But maintenance costs can rise occasionally in time and it is common for hydro-electric plant machinery to be renewed after 50–60 years to extend life. But machinery renewal costs are a small percentage compared to the overall capital costs of the dam's civil engineering structure. So as a general principle LCOE falls rapidly with age;
- Nuclear electricity:
  LCOE is clearly high, although not as high as renewables were in their earlier years. There is also a wide range of LCOEs in these reporting years, probably due to higher costs in smaller rather than larger nuclear plants.
  These figures are the result of the highly regulated nature of nuclear plant management. But the figures compare well with fossil fuel plant and nuclear has delivered consistent performance over many years.
  Another feature of nuclear is that, because of its high regulatory and construction costs, new nuclear plants are now being designed with much longer lives and, like hydro-electricity, these can and will deliver predictable, consistent performance in the future.

Figure 11.5 shows an alternative analysis of LCOE in \$/MWh and graphical form, between 2014 and 2021, drawn from consultant data, over a wider range of fuel sources than Figures 11.3 and 11.4 or Table 11.1.

This data demonstrates several interesting features, as follows:

- Photo-voltaic solar:
  Generally low LCOEs but with a wide range from different country sources, because of varying insolation;
- Onshore wind:
  Relatively low LCOEs at the lower end of energy costs;

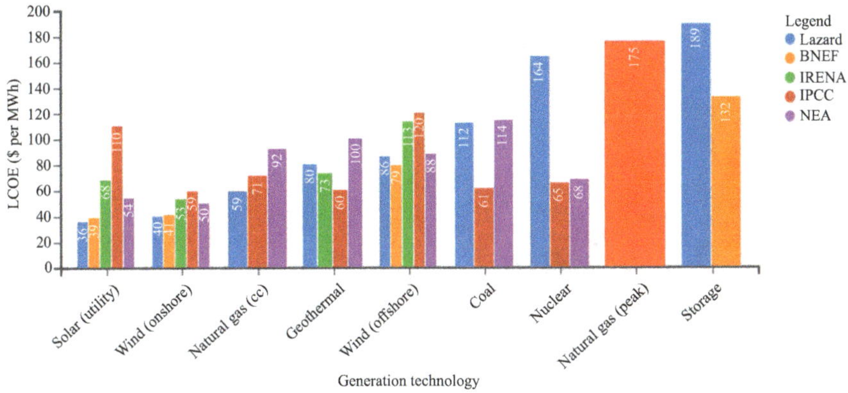

Figure 11.5 Various consultants' evaluations of global LCOE in $/MWh, from a wide range of energy sources, coal, oil, gas, nuclear, renewables and storage, 2014–21

Table 11.1 Summary of US fuel capital costs in $/kWh, from a range of fuels coal, oil, gas, nuclear, renewables and others, 2019–20

Fuel source	Generation type	Capital cost ($/kWh)	Data year
Coal	Without SO_2 or NO_x controls	3,500	2019
	With SO_2 and NO_x controls	3,800	2019
Gas	Combined cycle gas turbine (CCGT)	1,000	2019
	Simple combustion turbine	710	2020
Wind	Onshore	1,600	2019
	Offshore	6,500	2019
Solar	PV (fixed)	830	2021
		1,800	2019
	PV (tracking)	860	2021
		2,000	2019
Hydro-electricity	Conventional	2,752	2020
Nuclear electricity	Advanced	6,000	2019
Others	Battery storage power	1,380	2020
	Geothermal	2,800	2019
	Fuel cells	7,200	2019

- Offshore wind:
  Higher LCOEs than onshore wind, because of greater capital costs, but lower LCOEs than coal;
- Natural gas:
  Generally low LCOEs but the peak is higher than from any other energy source, due to the attractiveness of natural gas because of its transportability by pipe and sea;

Table 11.2 Summary of European Union energy production LCOE in €/MWh, from a range of different energy sources, 2009–21

Fuel source		€/MWh												
		2009		2011		2012	2013		2015		2018		2021	
Coal	Lignite	46	65	45	100	–	38	53	29	84	46	80	104	153
	Hard	49	68	45	100	–	63	80	40	116	63	99	110	200
Natural gas	CCGT	57	67	40	75	93	75	98	53	168	78	100	78	131
	Bio-gas		–		–	126	135	215		–	101	147	72	173
Wind	Onshore	93	50	130	65	81	45	107	29	114	40	82	39	83
	Offshore	–	120	180	112	183	119	194	67	169	75	138	72	121
Photo-voltaic solar	Small-scale	–	–		137	203	98	142	–		72	115	58	80
	Large-scale	–	–		107	167	79	116	35	180	37	85	31	57
Hydro-electricity		–	–		–		–		22	108	–		–	
Nuclear electricity		50	60	100	70	105	–		36	84	–		–	

- Coal:
  Variety of LCOEs perhaps reflecting different extraction costs in different countries and their varying regulatory influences;
- Geothermal:
  Medium LCOEs for a resource that is only available in relatively few countries, e.g. Iceland and Italy;
- Nuclear:
  A wide range of LCOEs reflecting uncertainty in the industry, due to widely varying regulatory influences in different countries;
- Storage:
  High LCOE because of high capital cost and the fact that storage requires two functional iterations, generation and storage. But these may not reflect the true longevity of this type of infrastructure.

11.5 Summary

The chapter opened by highlighting the duration and historic uncertainty of technological developments experienced achieving reliable operational renewable prototypes.

Note the following development periods for previous technology:

- Over 130 years for steam locomotion technology;
- Over 100 years for flight technology.

Thus, technological uncertainty and time are essential and inevitable components of new energy technology developments. That includes clean energies, although experience has shown that development times do fall with increasing experience in the modern age.

The chapter has also summarised the capital and LCOE costs available to date for a range of renewable energy sources producing electricity, compared to existing conventional sources, in the following ways:

- Energy quantity.
  Figure 11.3 shows that conventional coal, gas, oil and nuclear fuel sources are compact or 'concentrated', whereas in general renewable sources, solar, wind and tidal are more 'distributed' than conventional energy sources but are capable of providing a significant resource volume;
- Energy costs:
  - Fuel: Table 11.1, 2010–19. Fuel costs for renewables are low compared to fossil fuels because the fuel is effectively free, whereas for nuclear fuel, the cost is high but only very small quantities are required;
  - Generation: Figure 11.4, 2010–19. Capital costs for renewables, solar, onshore and offshore wind, are falling as collective power production increases and capital costs fall. Capital costs for nuclear rise as regulatory pressures increase, but plant life is also increasing.
- Levelised cost of energy:
  - Various data, 2009–21, show that from all fuels, fossil, nuclear and renewables, LCOE remains within relatively narrow bands, Table 11.2;
  - However, LCOE for coal and gas has risen slowly, whereas for renewables, LCOE has fallen slightly with time;
  - Final graphic, 2014–21, from consultants clearly shows LCOE ranges over all technologies, Figure 11.5.

The important lessons from this data are as follows:

- Hydro-electricity is a very important clean energy generation contributor, particularly but not exclusively in countries with mountainous geography, such as China, Norway, Sweden, Germany, France and Italy.
  Capital plant costs are high but effectiveness and long-life are strong;
- Nuclear energy has an important part to play in clean energy generation.
  Plant capital costs are high, but their effectiveness and long-life are strong and nuclear safety issues can be over-stated;
- Renewable energy is already making a significant cost-effective contribution to energy production in the countries reviewed here;
  - Wind and solar power sources are currently significant renewable energy developments but they both depend upon weather conditions and location; Capital costs are reducing, operational life is increasing, and effectiveness is reasonable;
  - Wave and tidal sources are still under development. However, technology development times suggest that we are still in the early stages for these resources, dependent on successful technology and location.

It is entirely possible that important breakthroughs will be achieved within the next 20 years;

- Coal, oil and gas generation have been effective but polluting energy sources, and their thermodynamic efficiencies have progressively improved with time but are still limited to 35%–40%. Their contribution will have to reduce to meet clean energy standards;
- Combined-cycle gas turbine generation is a more effective yet still polluting energy source with a lower capital cost and higher efficiency than other fossil fuel sources, 55%–65%.

  Combined-cycle generation is likely to continue to be a valuable energy source for some time to come, but its contribution will need to reduce to meet cleaner energy standards.

Chapter 12
The future

12.1 Introduction

This book has shown so far that there is a promising future for clean energy technologies but that they take time to develop and be deployed on any international scale.

However, there are a wider range of technologies now available, some of which are working well and growing, some are in early phases of development, and some are complex, struggling to provide cost-effective and reliable power.

A current overview of present and upcoming future technologies shows that World human energy usage, at least from the supply of electricity, is steadily increasing and that the division between sources is as given below:

- Coal-burning technology efficiencies have reached a plateau of about 35% for large plant sizes, 3–6 GW, and operational performance improvements are now limited by measures to reduce pollution consequent upon coal-burning.
  World annual coal-fired electricity production is currently about 10,000 TWh;
- Gas-fired combined gas and steam turbine technology efficiency has risen steadily to 65% for medium plant sizes of 0.6–1.5 GW and gas-burn pollution is less than for coal-burn.
  World annual gas-fired electricity production is currently about 6,300 TWh;
- Hydro-supplied.
  Supplied largely from huge hydro-electric dams located in relatively few locations, particularly China, South America, Europe, North America and Russia.
  World annual hydro-supplied electricity production is currently about 4,326 TWh;
- Nuclear-fired.
  World annual nuclear-fired electricity production is currently about 2,600 TWh;
- Wind powered renewable energy has advanced enormously, both in terms of technology and installed capacity. Larger wind turbines provide a good performance. Onshore wind turbines sizes range from 1 to 10 MW, demonstrating Capacity Factors up to 20% in windy locations. Whereas large offshore wind turbines range from 10 to 20 MW in very large arrays and are proving to have higher Capacity Factors of 25%–30%.

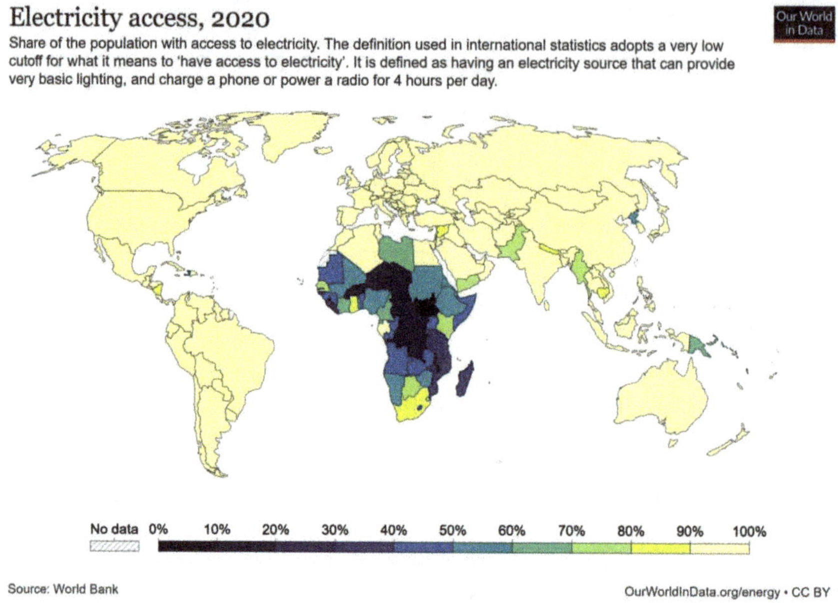

Figure 12.1 World-wide access to electricity, 2020

World annual wind-powered electricity production is currently about 2,139 TWh;

- Solar.
  World annual solar-supplied electricity production is currently about 1,289 TWh;

The future economic prospective effects of electrical energy and important variations across different continents are clearly shown in Figure 12.1, and the continent where the most improvement is needed is clearly Africa, the cradle of human civilisation.

It is still possible that there may be a future place for fossil fuels, in a better-balanced world energy economy, see [31], however, this is a contentious area.

A round up of the lessons to date about the use of fossil fuels and renewable sources, including technical lessons learnt enabling connectivity of distributed renewable energy sources.

Figure 12.2 shows in 2022 the current share of renewable energy sources being accessed around the World.

It should be noted that Norway, Canada, Brazil, Venezuela, Ecuador, Peru and Chile all have significant access to renewables, this is largely due to hydro-electricity.

But another striking feature is how low the access to renewables is in Africa, where electrical development has also been low.

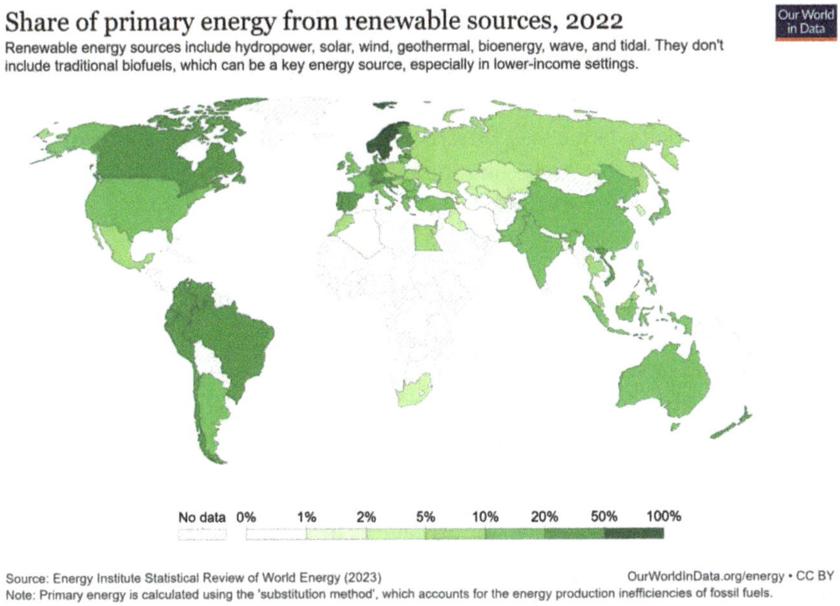

Figure 12.2 Share of primary energy from renewable resources, 2022

12.2 Clean energy technology futures

12.2.1 Trends in energy costs

By the middle of the twentieth century energy costs were falling, partly because of huge successive investments in coal, oil and gas exploitation.

Generally, this was happening in the developed World of Europe, Russia, North America and East Asia.

This was also facilitated in developed countries by top-down extensive National Electrical Grids spanning many hundreds and even thousands of km, fed by large thermal generation plants up to 3 GW, mostly burning fossil fuels, of relatively high thermal efficiency, up to 40%, supported in some countries by relatively few very large hydro-electric plants, also connected to the Electrical Grid. These plants but were also.

These works were largely undertaken through nationally organised campaigns, regardless of local politics, countries of the hard left and hard right industrialised power in very similar ways, see Giddens [22]. Compare the developments in the Soviet Union, United States and China.

All in the more developed countries designed and developed their energy systems in similar ways to improve national output and electricity prices were held low.

However, as industrial production rose, so electrical systems became more essential. Complexity and innovation demonstrated that larger numbers of bottom–up smaller size power generation plants with higher efficiencies could prove more cost effective to construct and extend energy supplies to a wider range of customers.

Access to electrical grids has improved but in the twenty-first century electricity prices have then started to rise and that has largely been the case world-wide.

Rising electricity usage has therefore led to the wider direct use of other, cheaper and transportable fuels, the prime fuels initially being coal, oil and gas. Alternative sources were as below.

12.2.2 Electricity

Electrical energy is at source a costly energy vector, because of inefficiency of conversion from fuel to electricity and generation costs, there are also losses in transmission and usage.

However, electricity has proven over 140 years to be very versatile, adaptable to a wide range of users, from local householders to local authorities and businesses and to internationally giant industries. Therefore, we tolerate the costs because of its flexibility and it continues in usage all over the World.

The only area that is somewhat resistant to the spread of electricity is Africa and that is because of the enormous sizes of African countries, the high costs of maintaining electrical plant and transmitting energy and the lack of African nations capital resources to build the power stations and transmission infrastructure.

12.2.3 Water

The availability of water is hugely important to both developed and developing economies.

Those countries with large rivers, such as China, Canada, Russia and Brazil, have derived tremendous electrical, industrial and agricultural benefits from their water resources.

Africa has extensive rivers, but with lower flow rates, and has generally been unable to exploit the full potential of the hydro-electric resource and benefit from it, except in Egypt, where the Aswan High Dam generates electricity for the whole country and the Aswan Lake irrigates large parts of Northern Egypt.

The causes are that African river flow rates do not generally compete with the highest flows in other continents and there is, so far, a lack of African capital to make these investments possible.

12.2.4 Air

At the Earth's surface air has a density of approximately 1.204 kg/m^3 at a pressure of 101 kPa and temperature of 20 °C.

Compressed air has proved valuable for many tasks and operations, such as:

- An important medium for energy transfer in industrial processes;
- Automation by the operation of air cylinders to actuate successive processes;
- Raising and lowering lifts;
- Small road vehicle propulsion;

- Operating the doors and air brakes for large main-line trains and the underground;
- Air brakes are also found on large highway vehicles;
- Power tools such as air hammers, drills, wrenches and paint atomization;
- Supply of breathing gas for underwater divers.

But compressed air lacks the power of high energy fuels, such as coal-gas, gas or oil, to play a major part in World human energy needs.

However, none can contend that it is not supremely valuable for the work it does in sustaining life and that is where, technologically we should leave it.

So, while its industrial application is widespread its energy contribution is limited because of its lack of power.

12.2.5 Hydrogen

12.2.5.1 Introduction

Hydrogen is a colourless, odourless gas with the lowest density of all gases and has been seen, by some, as a clean fuel of the future.

But it has not been used as domestic fuel, due to several reasons hydrogen is not easily available, and cost of production can be high.

Why hydrogen has not been used as a fuel is because, unlike other gases, it is not readily available in the atmosphere.

However, hydrogen is set to play a significant future role in the energy field because of its physical properties.

It requires processes like the electrolysis of water for its production, which is a costly and time-consuming process. However, it can be manufactured by steam reforming briefly described below.

However, there are some significant but not insurmountable barriers to its wider use, in particular:

- It is a low density as a gas, which means it requires very secure storage vessels and pipelines, much more carefully sealed than oil or natural gas storage tanks or pipelines;
- Hydrogen gas can be pressurised, liquefied and stored in high pressure steel vessels or pipes, the daily task of industries like the British Oxygen Co, Linde or Air Liquide.
- These technologies are already in use as hydrogen is a widely used coolant for electrical generators in large power stations, where the hydrogen is manufactured, liquefied and stored on-site until required for use;
- Its flammability, which means that special care must be taken to seal low- or high-pressure hydrogen storage vessels and pipelines to prevent leakage and possible ignition;

12.2.5.2 Blue hydrogen

Blue hydrogen is produced mainly from natural gas, using a process called steam reforming, which brings together natural gas and heated water in the form of steam.

Blue hydrogen must incorporate carbon capture, utilization and storage (CCUS) into the process.

The output is hydrogen, but carbon dioxide is also produced as a by-product, so blue hydrogen process plants must also have customers for this by-product.

The blue hydrogen method has been the route to provide relatively low-cost hydrogen to some early plants looking to provide low-carbon energy and this is becoming available now.

12.2.5.3 Green hydrogen

Green hydrogen can be made from water via electrolysis and the process must be powered by a source like renewable energy to qualify as a low-carbon process.

The barriers to growth of green hydrogen are associated with the current high costs of electrolysis and the availability of the considerable wind, solar or biomass renewable energy input resources necessary to power the process, although some pilot plants have been operational.

It will be some years yet before green hydrogen can play a part in human clean energy production and it is likely that its application will for very specific high-tech applications.

12.3 Future nuclear power

12.3.1 Modular fission power

The high costs of large nuclear power stations are due, in part, to increasing regulatory constraints, because of the four incidents described above, but also due to the huge size of these plants.

Costs have also risen due to the high civil engineering costs for the preparations needed for building these projects on nuclear sites, the intricacy of large modern high-performance nuclear equipment and the regulations. This has then led to extended construction times.

However, that has been mitigated by much longer prospective nuclear plant lives, now set at 60 years or more.

High capital costs have forced nuclear plant manufacturers and operators to consider other, lower cost alternatives.

This is why there is now a move to develop smaller, modular mass-produced reactors, which will have lower outputs with smaller constructional footprints, but lower civil engineering costs and easier to install reactor systems, see Figure 12.3.

12.3.2 Fusion power

A few projects are underway world-wide to develop nuclear fusion reactors in which where light element atoms, such as hydrogen, are joined together under neutron bombardment in a reactor, releasing fusion energy. Fusion reactions are more energetic than fission reactions.

Fusion power has been proposed as a form of power generation generating electricity by using heat from nuclear fusion reactions. In a fusion process, two

Figure 12.3 Example of a proposed 470 MW unit UK Modular Nuclear Reactor. Courtesy: Rolls Royce

lighter atomic nuclei combine to form a heavier nucleus, while releasing energy. Devices designed to harness this energy are known as fusion reactors. Research into fusion reactors began in the 1940s, but as of 2023, no device has reached net power.

Fusion processes require fuel and a confined environment with sufficient temperature, pressure and confinement time to create a plasma in which nuclear fusion can occur.

A combination of these conditions results in a power-producing system is known as the Lawson criterion. In stars, the most common fuel is hydrogen and gravity ensures extremely long confinement times that reach the conditions needed for fusion energy production.

Proposed fusion reactors use heavy hydrogen isotopes, such as deuterium or tritium, especially a mixture of these two, which react more easily than protium, the most common hydrogen isotope, allowing them to reach the Lawson criterion requirement under less extreme pressure and temperature conditions for the fusion reactor containment vessel.

Most fusion reactor designs aim to heat their fuel to around 100 million degrees, presenting major material and control challenges in producing successful reactor designs, Figure 12.4.

As a source of power, nuclear fusion has several potential advantages compared to fission, including:

- Reduced operational radioactivity;
- Less high-level nuclear waste;
- Ample supplies of hydrogen fuel are available with increased safety.

Figure 12.4 Example of an experimental fusion plant, Joint European Torus Tokamak magnetic fusion plant in UK, 1991

However, the necessary combinations of temperature, pressure and duration have proved difficult to achieve in a practical, economical manner. Another issue affecting this process is the management of neutrons released during the reaction, which will over time degrade many of the commonly used reaction chamber construction materials.

Fusion researchers have investigated various confinement concepts.

The early emphasis was on three main systems:

- The Z-pinch;
- The stellarator;
- The magnetic mirror.

The current leading designs are the Tokamak and Inertial Confinement (ICF) by laser.

Both designs are under research at very large scales, most notably at the ITER tokamak in France and the National Ignition Facility (NIF) laser in the United States. Researchers are also studying other designs, which may offer cheaper approaches, including an increasing interest in magnetized target fusion, inertial electrostatic confinement and new variants of the stellarator.

12.4 Networks

AC electrical systems are widespread throughout the developed World, the largest and most highly interconnected EHV AC system being in Russia, interconnecting the countries of the former Soviet Union.

The various voltages described in the following section are defined as they appear in the abbreviations:

- Extra high voltage (EHV), 345–765 kV AC;
- High voltage (HV), 100–345 kV AC;
- Low voltage (LV), 100–1000 V AC;
- Medium voltage (MV), 1–100 kV AC;
- High Voltage Direct Current (HVDC) supply, between 100–800 kV DC.

EHV and HV DC interconnection schemes have also grown either for:

- Long distance, large EHV DC energy transfer volumes from remote hydro-electric plants to large EHV AC substations near centres of population, for example in Canada and China;
- Medium distance HV DC transfer of energy from offshore wind farms in Northern Europe to EHV or HV AC substations near centres of population onshore;
- Or short back-to-back HV DC interconnectors have been widely employed, particularly in Northern Europe, around the Baltic and North Seas and in Japan and China, to couple AC systems of similar or slightly differing frequencies, providing a national integrated energy system.

In large territories or extensive mountainous region, for example in Africa or Central Asia large-scale AC networks have proven difficult to develop.

This has been because the long distances involved between centres of human activity, high capital costs and low economic outputs, which have prevented governments from investing in what can prove to be very costly projects.

There have been some large-scale international investment interventions to encourage such schemes, and this has been successful in South America.

In Africa, large electricity scheme investment levels have been low, apart from in the north and far south, where economic activity has stimulated electrical growth.

As a result, there is some discussion now of the development of bottom–up, low energy electrical growth, based upon the Internet, satellite links and mobile phone communications have been suggested for areas like Central Africa and Central Asia.

But significant technical and investment developments are necessary to make such schemes viable so that they could deliver meaningful levels of power.

12.5 Storage

12.5.1 Advanced gas networks

12.5.1.1 Air

Pressurised air networks are present in some large industrial plants, particularly in the automotive industries in the US, but they are not widespread in modern industries in other countries.

There could be a revival of this type of cheap energy distribution in developing countries for powering large numbers of hand tools.

12.5.1.2 Hydrogen

There has been considerable interest in developing the use of hydrogen as a gas with significant energy content to power distributed plant in large factories, because of the easy availability of the element.

But there have been concerns about the safety of such distribution systems.

12.5.1.3 Gas storage

There has been considerable innovation in the process of gas storage.

Some have advocated storage at the Earth's surface in gasometers, as was done to store coal gas, Figure 12.5(a).

However, the oil and gas industry has made use of underground stores in the form of depleted oil and gas fields, salt caverns or saline aquifers, Figure 12.5(b). However, such methods would not be suitable for the storage of low density hydrogen, which would require much higher integrity enclosures.

12.5.2 Advanced electricity networks and developments

Europe, the US, South America, Russia, India, China and Japan all have highly developed, extensive AC electrical grids that have grown from the end of the nineteenth century to the present day, with a wide range of EHV and HV transformers, AC voltages, with sophisticated switchgear, highly organised control systems and centralised grid controls.

However, modern electronic control technology is still advancing and many of these networks are being operated in ways that have changed dramatically over the past 20 years, because of the impact of detailed data information, computer forecasting and computer control.

12.5.2.1 HVAC

The World's widespread availability of electrical generation plants has expanded the extent of our HVAC networks all over the World with the single notable exception of the centre of West, Central and Eastern Africa.

12.5.2.2 HVDC

HVDC links have developed strongly, for example in Europe and China, because of the need to interconnect much larger AC networks operating at nominally the same frequencies.

(a)

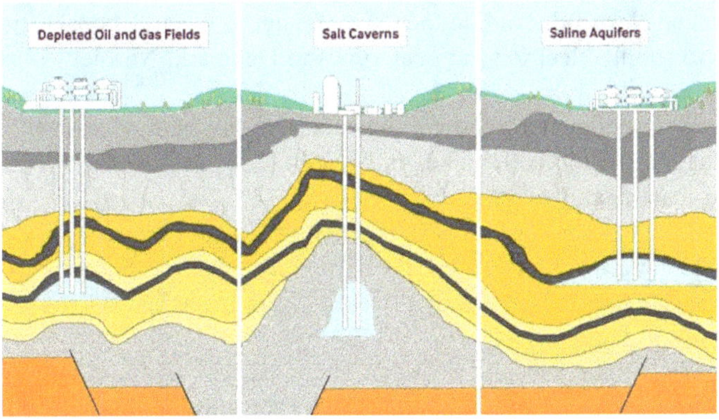

(b)

Figure 12.5 Gas storage examples. (a) Example of a UK elevating gasometer, storing coal gas and (b) underground natural gas storage

Occasionally HVDC has been used to couple HVAC systems operating at differing frequencies, e.g. in Japan.

HVDC links have been used recently to connect remote, large offshore wind assets, to onshore and reduce transmission losses using HVDC, to overcome the higher losses that would accrue if HVAC were used.

12.5.3 Electrical storage

12.5.3.1 Network storage and batteries

Battery storage has long been an important but small part of the operation of electrical power systems.

Initially, it used for low-storage capability, lead-acid batteries to provide low level back-up power to circuit breakers and for starting standby diesel-power for power station black-start facilities.

Improved technology, more compact and powerful lithium-ion batteries, with greater storage potential, have been developed for use in electric cars and aircraft, have become capable of more significant roles in the power electrical industry and larger interventions are now possible.

There have been a number of attempts to use much larger flow-batteries and converter power electronics to support the Grid but in UK these experiments have not proved successful.

12.5.3.2 Hydraulic storage

Hydraulic storage has been exploited by large national electricity grids, with hydro-electric facilities, to balance power requirements at times of peak demands.

This practice has increased considerably since the 1960s as electrical grids expanded and grid operators can rely upon large hydro-electric reservoir operators to hold back energy and release it to meet grid needs.

In Europe, Norway's and France's large hydro-electric capabilities can be relied upon to supply electricity at peak times to Denmark, Northern Germany and Italy, respectively.

However, large hydro-electrical installations are not necessarily required. Great Britain and Ireland both make good use of their much smaller hydro-electric resources to meet peak demands.

In Ireland this is done by adjusting the River Shannon flow rate for the hydro-electric plants there, where river levels can be retained when electric power supply is adequate but increased flows can be released from the hydro-electric plants when electrical demand is greater.

In Great Britain, Wales and Scotland, there are four hydro-electric power stations that can retain water and pump-back into their hydraulic reservoirs for storage, later releasing energy at times of high-power demand. The pump-up is sometimes powered by excess wind power from nearby wind turbines and that energy stored hydraulically and can released from the hydro plant when power demand is high.

These capabilities are facilitated nowadays by high-speed grid communications, well-documented long-term meteorological and engineering experience of the retained energy in their hydraulic basins and now by the modern use of big data and AI.

12.5.4 Future developments

In the future it is likely that traditional top-down electrical energy supply will change, particularly in more economically developed countries and this is already happening in the wider European electrical network, Figures 7.8 and 7.9.

One aspect of electrical energy, which is starting to occur in developing countries, is the bottom–up supply of electrical energy and this particularly noticeable in Central, East and West Africa, where the costs of installing expensive nation-wide transmission grids are currently prohibitive.

Figure 12.6 *Tabitha Goldstaub, Proponent of artificial intelligence's future importance to society*

12.5.5 Advanced networks, control and storage for energy despatch

The issue of bottom–up energy supply also emphasises the importance of artificial intelligence (AI) for the future management of energy grids and systems.

Tabitha Goldstraub, Figure 12.6, is a founder of CognitionX and an exponent of the importance of AI in the management of human data and an advocate for the potential of AI to improve markets and societies, as well as an activist for many of the unanswered ethical considerations around the increasing use of AI.

CognitionX is a market intelligence platform for all things AI, whose mission is to bring clarity to this complex and fast-paced world, to help ensure responsible transitions towards an increasingly AI-driven society by offering insights and independent advice.

Tabitha is especially focused on raising people's consciousness regarding the risks of building existing bias and sexism into AI.

12.6 Clean energy economics

12.6.1 Increased capital costs of distribution

The distribution of energy, whether through oil, gas pipelines or electrical networks, is costly.

Oil, gas and water pipeline costs per km have gradually fallen with time, the development of pre-installed components and the use of on-site welding and improved transportation.

EHV electrical transmission costs have also reduced per km with the development of lower cost towers and use of advanced stringing technologies.

However, pipeline and electrical MV and LV distribution costs within populous towns and cities with dense populations have steadily increased.

The impact of laying pipelines, cables and overhead lines in busy city industrial and residential areas, made more complex by roads, housing and infrastructure, has become increasingly problematic.

However, there are technological instrumentation, data management and computer control techniques that are becoming available which can and will reduce these difficulties.

12.6.2 Reducing pollution

The detailed task of reducing pollution lies with the operators of our energy producing and industrial economy plants but also with our society, which defines the standards that must be worked to.

However, the economy is also influenced by the competition between countries and our need to compete with others on an equal footing. The formation of trading relationships, for example the European Union (EU), has encouraged the imposition of common standards and that has certainly been valuable in Europe for placing competing countries within the Union on common pollution standards.

It is unclear what effect the UK's leaving the EU will have on the standards being applied in the energy plants of the UK.

It is interesting that when coal-fired power stations were dominant in the UK, that other EU countries, particularly Denmark and Sweden, were vociferous in encouraging the UK to moderate the pollution being emitted by the UK coal-fired power stations, generally towards the NW, because of the effects of that pollution on those countries own forests and environment.

12.6.3 Applications by geographical area

12.6.3.1 Europe

It is interesting to observe the differences between different countries in their standards for installing urban electricity, as follows:

- In UK, we have a tradition of laying MV and LV distribution cables underground in conduits or trenches and then bringing supplies up into buildings and houses. The trenching and cable box systems used to be standardised but as electricity supply has expanded are now frequently rather chaotic and one suspects costly;
- In Italy, with in towns and cities, many more ancient traditional churches, buildings and tenements, the tradition has been to supply MV and LV electricity via aerial cables with sophisticated, routing and connector systems

packaged in elegantly designed, standardised containers. The effect is organised and non-intrusive, but probably cheaper than the UK solution.

12.6.3.2 North America

North America has an integrated HVAC electrical network consisting of two alternative approaches from the US and Canada:

- United States: Developed an economically evolved HVAC, 60 Hz electrical network, as originally conceived by Westinghouse and Tesla, in contradiction to Edison. That network has expanded organically, largely under the effects of competitive capitalism. But it was also influenced by the 1920s depression and the shock treatment of Roosevelt's New Deal, which developed the Public Works Administration programme from 1935 to 1944, including the building of large hydro power stations with interconnected HVAC transmission systems of 190,000 km. These have stood the USA in good stead for many years;
- Canada: After the Second World War the Government developed a programme of electrical works, loosely based on Roosevelt's New Deal programme.

It included extension of its 60 Hz transmission network to over 160,000 km. These electrical grids follow north-south orientations, as Canadian population centres are concentrated along the US border, while large hydro-electric and nuclear projects are located to the north, in sparsely inhabited areas. This situation forced Canadian utilities to innovate. In 1965 Hydro-Québec commissioned its first 735-kV AC link in the Manic-Outardes project to the Lévis substation. In 1972 Manitoba Hydro connected generating stations on the Nelson River Hydroelectric Project to Winnipeg through an HVDC power line, the Nelson River Bipole.

There is strong electrical and trade integration between Canada's provinces and the US, so the Canadian transmission network is largely integrated with the US network;

- An amusing final vignette arose on this subject, at a London IET meeting which the author attended, reviewing the US's Northeast 2003 blackout.

An American in the audience suggested the incident may have been triggered by Canadian transmission system faults, at which there was much audience laughter with the American visitor asking, why? It was explained that it had been widely understood in UK that the Canadian part of the transmission system was very reliable but that US parts were less robust. The blackout's cause was subsequently pinpointed to a software bug in FirstEnergy's Akron, Ohio company Control Room alarm system, rendering operators unaware of a need to redistribute load after overloaded transmission lines in New York state had drooped into tree foliage.

12.6.3.3 China

China has a strong and extensive 50 Hz AC grid, but like North America it has evolved effectively from three large AC networks, in the north centred on

Beijing, in the centre based on Shanghai and the Yangtze River basin and in the South based on Hong Kong and Guanzhou. These three separate 50 Hz networks are not synchronised but are each interconnected by a number of back-to-back HVDC links.

12.6.3.4 Africa

Africa has no continent-wide Electrical Grid; however, there is an extensive grid across the whole of the north African coastline, in Egypt down the Nile River, and in South Africa and adjacent states. It is difficult for inland parts of Africa to justify the cost of building an Africa-wide Electrical Grid, but it would certainly be beneficial for economic development.

12.7 Summary

The concept of Clean Energy has been a human aim for energy production for more than two centuries.

But human technologies have been unable to facilitate the application of Clean Energy solutions, until the most recent 100 years.

One could also argue that we are only going to be able to implement Clean Energy by the application of the most recent technologies, in particular our ability to control large distribution systems by the application of distributed electrical controls, wired/wireless communications and perhaps AI.

The lessons of the last 200 years are that technology has enabled human beings to effect global warming, but that the science that has thereby been developed from that process could deliver us from its effects.

Climate change knowledge has initiated different social and technological effects in human society.

On the one hand a political movement that simply says *'Just Stop Oil'*, on the other a scientific movement that tries to mitigate the effects of Climate Change, by understanding its causes and reducing or even eliminating its effects.

This book has been about the latter.

Chapter 13

Overall summary and conclusions

13.1 Fossil fuels

Fossil fuels are energy-dense but a finite resource.

However, following the Industrial Revolution they were the easiest and cheapest fuels to extract from the Earth and became very widely used.

The technology of offshore oil and gas exploration, expanding in Developed World countries with strong economies, has also opened up the opportunity to exploit those widely available offshore energy resources in the Third World.

This has become an important energy driver, particularly for poorer countries with significant offshore oil and gas assets, enabling them to industrialise, earn money and improve the quality of life for their population.

Advances in fossil-fuel fired thermal plant sizes have seen steady efficiency improvements, from 20% to 40% over the past 80 years, with a growth in unit electrical unit output size up to 1 GW, arranged now in power stations plant sizes ranging up to 6 GW.

However, in recent years some of those efficiency improvements have changed and even been eroded, from 30% to 36%, by the need for remedial post-boiler exhaust methods to reduce pollution.

The jewel in the crown of fossil-fuel fired thermal plants has been the development of combined cycle gas turbine generators using gas as their fuel and employing a combination of both a high-performance, advanced technology gas turbine followed by a high-performance steam turbine. These plants can achieve thermal efficiencies up to 65%, albeit in smaller plant size ratings of up to 1 GW.

Fossil fuels became a go-to source for energy expansion in each country progressing through its Industrial Revolution, from Great Britain to the United States to China.

But the use of fossil fuels for industrialisation, on a huge scale, has caused substantial environmental damage.

Present development trends for fossil fuels are increasing the exploitation of offshore energy sources, which has the potential to be lower costs and reduce international political impact.

However, the consequential environmental damage of burning these fuels will continue to be significant and tend to increase as more developing countries industrialise.

It is interesting that one of the most active areas where decarbonisation is taking place is in transport, electric cars and trucks have achieved significant sales and are driving improvements in electric propulsion of vehicles.

Therefore, such developments could merely prolong greenhouse gas emissions while despoiling the ocean bed exploiting offshore oil a gas fields, as humans had previously despoiled terra-firma by exploiting coal.

In 2021 the electrical energy generated by these fossil resources world-wide was 17,000 TWh.

13.2 Renewables

Renewable technologies in solar and wind power generation have seen enormous performance improvement over the last 50 years and these technologies, now support onshore wind and solar farms with power outputs up to 1 GW with an installed world-wide capacity of about 906 GW.

In 2021 the electrical energy generated world-wide by these renewable resources was 8,300 TWh.

Offshore wind power has developed particularly rapidly, largely because onshore wind experience had created a platform for offshore expansion with high energy resources available in the seas around Northern Europe and China, so that offshore wind farms with outputs more than 2 GW are now in development.

Other offshore renewables technologies, such as wave and tidal power, are still in their development phases.

There are many prototypes in operation in Europe, North America and East Asia. Operating experiences have proved challenging, but prototypes are evolving.

It is very likely that within the next 20 years reliable offshore wave and tidal devices and arrays of significant GW size, see Tavner [28], could be successfully available.

13.3 Nuclear

Nuclear fission power has continued to play a small but significant part in the World energy industry as an electricity generator, particularly where fossil fuels are sparse.

In 2021 the world-wide low-emission dispatchable electricity generated by nuclear power reactors was 2,653 TWh, accounting for about 10% of total global electricity generation and more than a quarter of the world's low-carbon electricity generation.

World-wide 25 countries currently operate nuclear power stations.

There is potential for clean nuclear fusion power and several development projects are underway around the World, but the technology is complex and costly, so the development of reliable nuclear fusion power stations is still some years away.

13.4 Power networks

The developments of large scale, interconnected, electrical power networks, first developed in the Soviet Union, North America and Europe, have now become operational on all continents.

Globally 175–275 km of power transmission lines transmit each TWh of electrical energy per year.

It is also clear, however, that less economically developed regions, for example in Africa and Southeast Asia, have less well-developed interconnected electrical power networks, because of the large capital costs involved.

However, significant development of lower cost Internet technologies associated with lower power networks may prove more economical, affordable and effective for their types of economies.

13.4.1 Top-down high power networks

Top-down energy networks versus bottom-up energy networks.

Energy delivered by National Electrical Grids is a top-down energy network, industrial expansion in all large economies, from Europe to the Americas to India and the Far East, from the end of the nineteenth century to the present-day has been largely driven by top-down energy networks, exemplified in electricity supply by the following Table 13.1.

For one country, the UK, a summary of recent history of CO_2 emissions and electricity usage are shown in Figure 13.1.

It can be seen that emissions have fallen over a period of 30 years, as a result of reductions in fossil-fuel firing of power stations, while electricity consumption has also fallen slightly over the last 15 years.

13.4.2 Bottom-up low power networks

It is likely, in the future, that economic development will be become more dependent on lower power, bottom-up energy networks, rather than present reliance, at least in developed countries, on top-down, energy networks developed in the twentieth century, see Section 13.5.

Table 13.1 Summary of World fuel usage for electricity and electrical network lengths

Sources, 2021	**TWh/ annum**	**Network, 2021**	
Fossil fuels	17,000	Network length, km/TWh	175–275
Renewables	8,300		
Nuclear	2,653	Estimated range of World Electric Network Lengths, km	4,891,775–7,687,075
Total electricity generated world-wide, TWh	27,953		

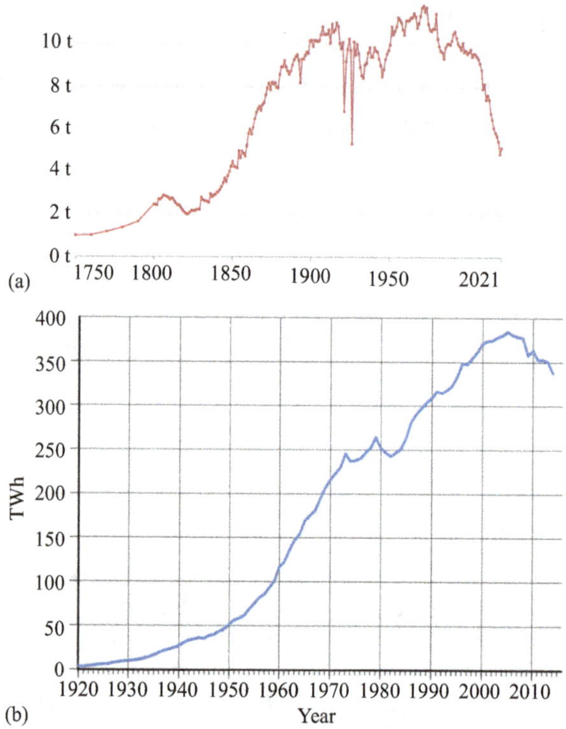

Figure 13.1 *UK CO_2 emissions and electricity usage. (a) 1750–2020, Annual per Capita CO_2 Emissions in tonnes, UK and (b) 1920–2015, annual electrical usage*

The widespread use of mobile phone networks, the development of Wi-Fi networks and the extension of these networks from cities into the countryside are increasing the use of voice and messaging communication, through the availability of low-power battery supplies and solar charging, with a reduction in the use of network power.

Gradually such low-power networks may become more important for controlling our high power electrical, communications and industrial networks, probably leading to a decrease in overall energy demand for power circuits.

There is also a particular case for wider deployment of low-power networks in developing countries where large-scale Electrical Grids are either prohibitively expensive, such as in remote mountainous countries in Asia, or more difficult to install, as in larger African countries with more remote interiors.

13.5 Storage

In 2020 total World installed hydro-electric energy storage capability was around 8,500 GWh that is on a base of 138 GW installed capacity by that year. This accounts for over 90% of global electrical storage, the USA having the largest capacity.

In the future, when utilising clean energy, storage is likely to be of increased importance both for the supply of high and low-power energy, increasing the reliability of both power and communication networks.

There is much current talk about the use of compact lithium-ion batteries, as these were developed for powering mobile phones, then electric cars and for use in lightweight aircraft.

Hydrogen may emerge to have potential as both an energy vector and storage source.

However, it is likely that lower cost storage elements, not using rare earth materials or relying on hydrogen, may emerge from new research, changing the face of low-cost storage, particularly for lower power networks. These developments may arise as connectivity moves gradually from high power needs towards the lower power needs for increased communication.

13.6 Clean energy, lessons learnt

An important conclusion from '*Clean Energy-Past to Future*' is to understand where there have been energy successes and where there have failures.

There has been the development of many different renewable energy techniques, ranging from hydro-electric power, through wind, wave, tidal, solar power and storage.

Of these technologies only hydro-electric power has fully matured, but the following indicates those technologies that are proven and those that are not yet.

13.6.1 Proven successes

Proven success has been human scientific progress, which has enabled us to differentiate between potentially successful or damaging energy sources.

Undoubtedly the two most significant scientific developments over the past 200 years, affecting our ability to deliver clean energy, have been:

1. Thermodynamics and heat engines
   This has helped us to differentiate between efficient and inefficient engineering processes, enabling us to concentrate on those areas of energy production which will be most efficient and least polluting.
2. Magnetism, electro-magnetism and electricity
   These have enabled us to develop methods for transmitting energy over huge distances, enabling energy to be moved between countries and continents with diminishing losses.

Hydro-electric power and storage are proven technologies but costly investments, however, with proven long-life potential.

Wind power has been a proven energy success onshore and even more so offshore, because of an energetic atmosphere there.

However, it is not yet clear that 3-blade horizontal axis turbines are necessarily the optimal engineering solution and there is likely to be refinements and maybe large changes in wind power's direction in the future, perhaps towards larger 2-blade horizontal axis or tethered kite designs, both of greater power unit output.

Solar power is a proven technology whose capital cost steadily decreases and greater integration into new modern housing and larger buildings will ensure its future effectiveness.

Tidal power and storage are also proven technologies and costly investments, and there may be relatively few geographical locations where it can prove to be overwhelmingly effective.

Nuclear fission power electricity generation has proven its efficacy, despite two serious accidents. There may be opportunities to improve costs, safety and longevity and reliability if modular nuclear fission technologies are developed.

Condition monitoring has improved the reliability of many renewable technologies, see Tavner *et al.* [28] and Tavner [34].

13.6.2 Not sure

Wave power is still promising and a potentially powerful technology, but it has not yet proven its reliability, repeatability or durability. The author believes that it will, but there is much work yet to be done.

The development of low energy storage.

Lithium-ion battery technology has figured large in delivering clean energy in transport, especially in the automotive field.

Whilst lithium-ion batteries are vital to small-scale devices like mobile phones, their large-scale use in electric vehicles and aircraft is still problematic and there maybe future problems for them fitted in lorries, trucks, buses and aircraft. This is an area of technology which needs to mature if it is to have a long-term clean energy future.

The development of bottom-up communication networks for distributing low levels of energy in remote locations is also an important potential for energy futures.

Nuclear fusion power electricity generation is still under intensive research in several countries. Its costs are high, but the potential benefits are also high. The author believes that this research should continue.

13.6.3 Failures

Important failing areas for modern energy are our present use of fossil fuels, which are finite in extent, polluting and facing extinction.

The modern developed World is still heavily dependent on burning of coal, oil and gas, in power stations with thermodynamic efficiencies ranging from only 25%–38%.

The burning of coal and oil to generate electricity with these low efficiencies is polluting and needs to end.

However, the burning of gas in combined cycle gas turbine power stations with higher efficiencies, closer to 65%, may continue for a period because of that efficiency but will need to finish within the present century.

The use of diesel and petrol in cars, lorries, trucks and buses is polluting and needs to end.

13.7 Conclusions

The most frequently asked question to those engaged in climate change issues is 'Are our human activities really damaging the Earth's environment?'

The simple answer is
'Yes, and we are not only affecting the Earth's atmosphere but are also changing the Earth's near surface sub-structure, the oceans around us and affecting the animals and plants who co-exist with us'.

But humans are not the sole polluters of the earth's atmosphere and environment as Figure 13.2 demonstrates, but our high energy life-styles have dramatically altered the Earth's environmental balance.

We need to find a way to improve our environment, not by panic, breast-beating, disinformation or anti-technology, but by concrete, practical and technically sound action as set out in Section 13.6.

On the positive side there is a growing international realisation from citizens, scientists, industrialists and politicians that we need to reduce the human impact on our environment.

There is also a slowing down in the numbers of humans being born, so that human civilisations are likely to decrease in numbers in future centuries.

The problem is which issues to tackle first, this author believes that reducing the use of fossil fuels and reduction of carbon emissions require our first attention.

These are not impossible dreams, humans have recovered from several catastrophic environmental situations, some described in this book, for example:

- The disorderly exploitation of coal over a period of 200–300 years, from the fifteenth to twentieth centuries, particularly in Europe, India, China and the USA, has damaged the health of millions of those working in the industry and affected those countries.
- However, on the other hand, it also stimulated economies around the World bringing scientific, educational and medical benefits.
- In some of those countries the coal industry has ceased, and some environments have been restored notably in the British Isles and France.
- But a curious enthusiasm for our coal-mining and the coal-burning past still remains in those communities, where hard physical labour was needed, family purses and larders were full, coal-fired industries flourished and where steam trains and steam ships hauled the majority of produce.
- The discovery, from the nineteenth to twentieth centuries, of prolific oil and gas resources in the USA, Romania and the Caucasus and their chaotic early exploitation, caused many casualties and deaths and serious consequential environmental damage.

Figure 13.2 *Causes and consequence of Earth's atmospheric pollution. (a) Human-cause atmospheric pollution; (b) natural-cause atmospheric pollution; and (c) consequences of atmospheric pollution, human or natural causes*

- However, as with the coal industry this has also further stimulated economies around the World bringing more scientific, social, educational and medical benefits.
- The operational methods of these enormously economically successful, but environmentally damaging, coal, oil and gas industries has improved human development but has had enormous environmental consequences.

- They have generated wealth, which has increased the education and health of the World's human population. But there have been environmental changes, that are still underway, and we cannot yet predict how those changes will develop or what will be their consequences will be, despite the economic benefits.
- Politicians all know that there are deleterious effects but, as yet, they do not have the methods or means to define or control their environmental effects.
- The nuclear industry grew world-wide from the 1950s but received two shocks following the Chernobyl and Fukushima disasters. Both caused death and life-changing damage to health from radiation, entailing the commitment of enormous national resources, in the former Soviet Union and Japan respectively, to cleaning-up the consequent damage, which could only be undertaken at National Government level.

However, Chernobyl was a much more serious nuclear incident than that at Fukushima, although the effect of the Tsunami, which triggered Fukushima, on the Japanese coast was even more challenging for Japan than the nuclear accident.

Both disasters have influenced the future course of nuclear power by imposing on the industry much more widely accepted safety norms.

Many lessons have been learnt in the nuclear fission industry and therefore it is subject to stringent regulation, both nationally and internationally. It remains to be seen how that will be managed in the future, whether the capital costs of nuclear fission power can be developed economically or whether modular nuclear fission power could be produced at lower unit costs, while still meeting high safety demands.

The biggest question mark in nuclear power is the future possibility of nuclear fusion power, which has considerable large potential, particularly for populous countries;

- Figure 13.3 shows, from the UK, a vignette over 14 years from 2009 to 2023, of what has been achieved up to 2023 to reduce carbon usage in the electricity supply system, lowering CO_2 emissions.

This demonstrates the substantial fall in coal burning and the rise of wind, solar and biomass power industry energy inputs, combined with the use of higher efficiency combined cycle gas-burning.

Collectively, these have reduced the carbon content of electricity production in UK over that period from 589 down to 148 gCO_2/kWh.

Similar results have been achieved in other developed countries.

Readers are also asked to note, however, that the UK now also relies substantially on electricity imported from its neighbours, notably from France (from nuclear power), Norway (from hydro-electric power) and the Netherlands (from wind power);

- Much of what needs to be done to convert to cleaner energy can be done by changing our economies and energy-dominated life-styles, which depend upon individual needs and decisions, that do not necessarily align with international, national or corporate needs.

However, we also need the nation and its businesses to protect us while supplying us with goods and services.

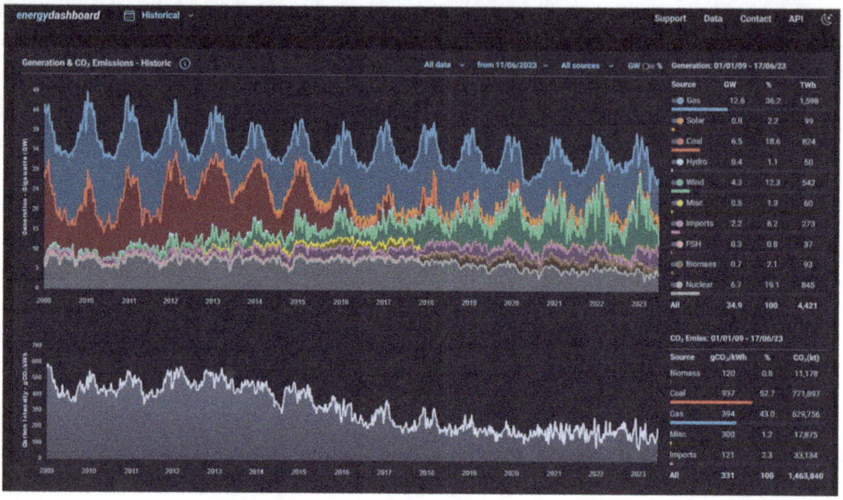

Figure 13.3 UK carbon emission reduction due to changes in electricity supply fuels, 2009–23, from Energy Dashboard. Net carbon reduction over 14 years, from 589 to 148 gCO_2/kWh.

Figure 13.4 The Eden Project, Bodelva, north-east of St Austell, Cornwall, UK

This is not necessarily intrusive, although it could be, but it does mean that Governments and businesses must adjust to accommodate public need.

This seems like a pipe-dream but businesses rely upon customers and Government support;

- In the UK a striking example of changing environmental attitudes was seen in Southeast Cornwall at the beginning of the twentieth century and is my last vignette.

Previously this part of Cornwall was an area exploited, during the nineteenth and early twentieth centuries, for the extraction of china-clay, providing valuable jobs and materials for the ceramics and other higher tech industries.

However, in an otherwise beautiful Cornish countryside, china-clay extraction created hideous lunar landscapes.

When, at Bodelva north-east of St Austell, extraction ended in the 1970s, it led to an exciting local Cornish initiative to improve that environment by developing the former china-clay mine, with no soil or plants, and transforming the huge crater into a beautiful garden of global interest, the Eden Project, which opened in 2001 (Figure 13.4).

Eden Project Co-Founder, Tim Smit said:

> 'It was ridiculous to imagine that the Eden Project was possible, and yet hundreds of people, trained to say NO, were persuaded to say YES!'

References

[1] White, G (1789) *The Natural History of Selbourne*, Penguin Classics, London.

[2] (a) Faraday, M (1839) *Experimental Researches in Electricity*, Vols. 1 and 2, Richard and John Edward Taylor, London; (b) Faraday, M (1855) *Experimental Researches in Electricity*, Vol. 3, Richard Taylor and William Francis, London.

[3] Smiles, S (1862) *Lives of the Engineers: With an Account of Their Principal Works*, John Murray, London.

[4] Jevons, W S (1865) *The Coal Question*, Macmillan, London.

[5] Maxwell, J C (1873) *A Treatise on Electricity and Magnetism*, Vols. I and II, Clarendon Press, Oxford.

[6] (a) Heaviside, O (1893) *Electromagnetic Theory*, Vol. 1, The Electrician Printing and Publishing Co., London; (b) Heaviside, O (1899) *Electromagnetic Theory*, Vol. 2, The Electrician Printing and Publishing Co., London; (c) Heaviside, O (1912) *Electromagnetic Theory*, Vol. 3, The Electrician Printing and Publishing Co., London.

[7] Carson, R (1951) *The Sea Around Us*, Oxford University Press, Oxford.

[8] Putnam, P C (1953) *Energy in the Future*, D Van Nostrand Co. Inc., Princeton, NJ.

[9] Sprague de Camp, L (1960) *The Ancient Engineers*, MIT Press, Cambridge, MA.

[10] Carson, R (1962) *Silent Spring*, Houghton Mifflin Mariner Books, Boston, MA.

[11] Needham, J (1994) *The Shorter Science and Civilisation in China*, Cambridge University Press, Cambridge.

[12] Cardwell, D S L (1994) *History of Technology*, Fontana Press, London.

[13] Greenhill, B and Morrison, J (1995) *The Archaeology of Boats and Ships*, Conway Maritime Press, London.

[14] Philander, S G (1998) *Is the Temperature Rising? The Uncertain Science of Global Warming*, Princeton University Press, Princeton, NJ.

[15] Scaife, W (2000) *From Galaxies to Turbines: Science, Technology and the Parsons, Family*, Taylor & Francis, London.

[16] Duffy, M C (2003) *Electric Railways, 1880–1990*, Institution of Energy Technology, Loughborough.

[17] Gleich, J (2004) *Isaac Newton*, Harper Perennial, London.

[18] Fortey, R (2005) *The Earth: An Intimate History*, Harper Perennial, London.

[19] Stern, N (2007) *The Economics of Climate Change: The Stern Review*, Cambridge University Press, Cambridge.

[20] Drury, W (2009) *The Control Techniques Drive and Controls Handbook*, 2nd Ed, Institution of Energy Technology, Loughborough.
[21] McKay, D J C (2009) *Sustainable Energy: Without the Hot Air*, Vols. 1 and 2 UIT, Cambridge.
[22] Giddens, A (2009) *The Politics of Climate Change*, Polity Press, Cambridge.
[23] Jackson, T (2009) *Prosperity Without Growth: Economics for a Finite Planet*, Earthscan, Oxford.
[24] Harari, Y N (2011) *Sapiens: A Brief History of Mankind*, Vintage, Penguin Random House, London.
[25] Allwood, J M and Cullen, J M (2012) *Sustainable Materials with Both Eyes Open*, UIT, Cambridge.
[26] Kolbert, E (2014) *The Sixth Extinction: An Unnatural History*, Henry Holt and Company, New York.
[27] Patterson, W (2015) *Electricity Vs Fire: The Fight for Our Future*, Walter Patterson, Glasgow.
[28] Tavner, P J (2017) *Wave and Tidal Generation Devices: Reliability & Availability*, Institution of Energy Technology, Loughborough.
[29] Rosling, H, Rosling, O, and Rosling Rőnnlund A (2018) *Factfulness*, Sceptre, imprint of Hodder & Stoughton, London.
[30] Tavner, P J, Ran, L, and Crabtree, C (2020) *Condition Monitoring of Rotating Electrical Machines*, 3rd Ed, Institution of Energy Technology, Loughborough.
[31] Yergin, D (2020) *The New Map: Energy, Climate and the Clash of Nations*, Penguin Press, New York.
[32] Jackson, N and Mecrow, B (2020) *Electrical Machines Road Map*, Automotive Council & Advanced Propulsion Centre, Coventry.
[33] Peake, S (2021) *Renewable Energy: Ten Short Lessons*, John Hopkins University Press, Baltimore, MD.
[34] Tavner, P J (2021) *Offshore Wind Power: Reliability, Availability and Maintenance*, 2nd Ed, Institution of Energy Technology, Loughborough.
[35] Fisher, K (2022) *A Pipeline Runs Through It*, Allen Lane, Penguin Random House, London.
[36] Johnson, K L, Gray N D, Stone, W, *et al.* (2022) A nation that rebuilds its soils rebuilds itself—an engineer's perspective, *Soil Security*, 7, pp 1–13
[37] Frankopan, P (2023) *The Earth Transformed: An Untold History*, Bloomsbury, London.

Appendix A
Earth's historical timelines

Appendix A.1 Geological: courtesy Prof Richard Fortey [18]

Million years before present	Eras (pre-human)	Major sub-divisions	Events	Features
4550	Pre Cambrian	Hadean		
3500		Archean		
2500		Proterozoic		
543	Lower Palaeozoic	Cambrian		Trilobites and other marine animals appear
505		Ordovician		Ancient Ocean, Lapetus, at its widest
438		Silurian		Caledonian Mountains at their zenith, colonization of the land
408	Upper Palaeozoic	Devonian		Fish and amphibia
362.5		Carboniferous	'Ice Age' coal swamps	
290		Permian	Pangea Continent	Great extinction
245	Mesozoic	Triassic		
208		Jurassic	Modern oceans widen	
145		Cretaceous	Chalk widely deposited	
65	Tertiary	Palaeocene	Mammal and bird diversity	Great Extinction, end of the Dinosaurs
56.5		Eocene		
35.4		Oligocene	Alpine evolution	
23.3		Miocene		
5.2		Pliocene		
1.64	Quaternary	Pleistocene	Ice Ages in Northern hemisphere greatly affecting specially adapted mammals	

Appendix A.2 Social: courtesy Prof Yuval Noah Harari [24]

Billion years before present	Important events (pre-human)
13.5	Matter and energy appear, the beginning of physics
	Atoms and molecules appear, the beginning of chemistry
4.5	Creation of planet Earth
3.8	Emergence of organisms on Earth, the beginning of Earth biology
0.006	Last common grandmother of humans and chimpanzees
0.01 to the Present	Holocene: Initiation of another Great Extinction
0.0025	Evolution of genus *Homo* in Africa, first stone tools
0.002	Humans spread from Africa to Eurasia, evolution of different human species
0.0005	*Neanderthals* evolve in Europe and the Middle East
0.0003	Daily usage by *Neanderthals* of fire
0.0002	*Homo sapiens* evolves in East Africa
0.00007	The Cognitive Revolution
	Emergence of fictive language
	H. sapiens spreads out of Africa, the beginning of human history
0.000045	*H. sapiens* settle Australia.
	Extinction of Australian megafauna
0.00003	Extinction of *Neanderthals*
0.000016	*H. sapiens* settle America, extinction of American megafauna
0.000013	Extinction of *Homo floresiensis*
	H. sapiens the only surviving human species
0.000012	Start of the Agricultural Revolution
	Domestication of plants and animals
	Permanent settlements
0.000005	First kingdoms, script and money
	Polytheistic religions
0.00000425	First Akkadian Empire of Sargon
0.0000025	Invention of coinage, universal transferable currencies
	Persian Empire, a universal political order for the benefit of all humans
	Buddhism in India, a universal truth to liberate all beings from suffering
0.0000020	Han Empire in China and coming of Chinese Buddhism
	Roman Empire in the Mediterranean
	Foundation of Christianity
0.0000014	Foundation of Islam
0.0000003	Beginning of the Scientific Revolution
	Humankind admits its ignorance and begins to acquire unprecedented knowledge
	Europeans begin to conquer the Oceans and the Americas
	Entire planet becomes a single historical arena
	Rise of Capitalism
0.0000002	The Industrial Revolution
	Families and communities are influenced by States and Market
	Mass extinction of plants and animals caused by human activity
	Compare to previous mass extinctions caused by natural environmental conditions
The Present	Nuclear weapons threaten human survival
	Humans transcend planet Earth's boundaries
	Human scientific activity means that some organisms are increasingly shaped by intelligent design rather than Natural Selection
	Effects of Global Warming

Appendix A.3 Technical: with thanks to Prof Bill Drury [20]

Years BCE/ AD up to the Common Era	**Origin, location, originators**	**Technologies including:** • **Prime movers;** • **Structures;** • **Machines;** • **Control;** • **Electronics;** • **Reliability**	**Photo, diagram or note**
		Early history	
200 BCE	Persia, vertical axis wind machines first used		
70 AD	Heron, Greece, Pneumatica, Development of a model reaction steam turbine There is debate whether Heron invented this model, or whether it was stimulated from different examples developed by Heron? The model is a first century turbine By the end of the nineteenth century, this device lead to steam and then gas turbine power, first for electrical power stations, then for ships and aircraft		
634–644 AD	Persia, Iran Practical windmills built in Sistan, Persia-Afghanistan border region by the Rashidun Caliph Umar for grinding grain and pumping water In 1963 50 of them were still operating in Neh, Iran Hau (2006)	Vertical axis wind turbines (VAWT), efficiency <10%, long vertical axle driven shafts, rectangular blades Enclosed by a 2-storey circular wall, millstones at the bottom, rotor at the top Rotor: spoked with 6–12 upright ribs, each covered with cloth to form separate sails. Photo, 1900	

(Continues)

Appendix A.3 (Continued)

c650–900	Arrival in Scotland, England, Wales and Ireland of people from Northern Europe	Sutton Hoo, Anglo-Saxon ship burial	
		Typical Danish Viking ship used to invade England and Ireland	
1090	China, Kaifeng, Henan Province (河南开封) Development by Su Sung (苏松) of the world's first water-driven mechanical clock, incorporating an escapement mechanism	This was an exemplar water-driven mechanical clock but dismantled and not repeated Published with permission from the Needham Research Institute	

(Continues)

Appendix A.3 (Continued)

1119	Netherlands	4-blade horizontal axis wind turbine (HAWT) post-mills for milling grain, draining water, sawing wood Easy to yaw, but support might have been an issue Post-mills dominated milling and pumping in Europe until the nineteenth century 
1219	China Cf, earlier reported Chinese turbines	Reference to a possible common origin for windmills in W China or Persia, see above, Needham (1994) and Zhang (2009) Sheng Ruozi refers to wind-mills in the 'Placid Retired Scholar', Yelü Chucai (耶律初) 1190–1244, prominent Jin and Yuan statesman after fall to the Mongols in 1234 The passage refers to a journey in 1219 to Turkestan, modern Xinjiang (新疆) and of Hechong Fu (和冲) to Samarkand, modern Uzbekistan. These VAWT wind-mills had luffable self-adjusting sails, in response to wind conditions as the wind-mill rotated Yancheng City, Jiangsu province (盐城市, 江苏) Recent reconstruction of traditional Chinese lightweight wood VAWT, for water pumping, fitted with vertically luffable sails, to adjust power, in the maritime tradition  Recent reconstruction of a small traditional Chinese lightweight wood HAWT, for water pumping Again, luffable sails 

(Continues)

Appendix A.3 (*Continued*)

1200s	Europe Cf, earlier 1119 Netherlands cap-mill	4-blade, squat HAWT structure, with wooden shutters adjustable as luffable blades
1295	Netherlands	4-blade HAWT upwind Cap-mill Advantage of cap-mill over post-mill is that it is unnecessary to turn the whole buck or mill body into the wind This allows more space in the tower for machinery and storage 
1424	China	Admiral Zheng He (鄭和) sailed this 8-mast ship from China to Palembang in Sumatra His ship is here compared to Columbus' 3-mast and bowsprit *SANTA MARIA*, 1492 
1700	UK	Watermill for milling flour 

(*Continues*)

Appendix A.3 (*Continued*)

1750–80	Abraham Darby, UK Beginnings of the European Industrial Revolution	UK's earliest established Iron Works at *Coalbrookdale*, Staffordshire for manufacturing iron, wrought iron and steel	
		Including manufacture of the components for the *Ironbridge* at nearby *Coalbrookdale*, designed by Darby	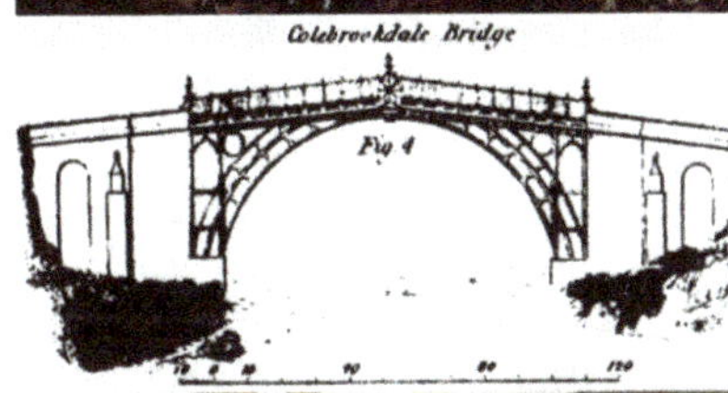
1776	James Watt, UK	Double Acting stationary steam engine for water pumping and draining mines Watt's design built by D. Napier & Son, London, 1832, Superior Technical School of Industrial Engineers, Madrid, Spain	
1800s	Leibnitz, Bernoulli, Euler, Netherlands and Germany	Development of precision wooden pin and socket gears for cap-mills	Patent for a precision right-angle wooden gearbox between horizontal axis mill rotor and vertical axis wallower gear, transferring horizontal to vertical rotary power Influenced by seventeenth and eighteenth century scientific innovations from Leibnitz, Bernoulli and Euler, based on the emerging mathematics of the natural sciences

(*Continues*)

Appendix A.3 (*Continued*)

1825	Robert Stephenson and Co, UK	*Locomotion No 1*, coal-fired locomotive for NE England early railway Design by Robert Stephenson based on his father George's experience, manufactured in Newcastle-upon-Tyne	
1829		*Rocket*, further developed coal-fired locomotive for the Rainhill trials on the Manchester to Liverpool railway Designed by Robert Stephenson with advice from George Stephenson and Henry Booth, manufactured in Newcastle-upon-Tyne	

(Continues)

Appendix A.3 (Continued)

Coal, oil and gas pollution sources combine		
Nineteenth century industrial activities in UK	 Industrialisation and coal burning, Black Country	 Railways
	 Lot's Road Power Station, London	 Steelworks, Sheffield

(Continues)

Appendix A.3 (Continued)

		Start of the electric age	
1820–73	Oersted, Faraday, Arago, Babbage, Herschel, Pixii, Lenz, Wheatstone, Cooke, Gramme, Pacinotti, Europe and USA	Development of the first DC electric generators and motors using current switching via rotating commutators	
1882	Edison, USA	Beginning of the electric age First central power station developed with large DC generators and underground LVDC cable for large-scale transmission of DC electric power from Pearl Street Power Station, central New York Efficiency <4%	
1885	Ferranti, UK, Subsequently, Tesla and Westinghouse, USA	Beginning of the electric age First central power station with AC generators, transformers and underground HVAC cables, from Deptford Power Station to Central London, for large-scale transmission and usage of AC electric power Efficiency <10%	 

(Continues)

Appendix A.3 (Continued)

		Cleaner energy the push-back starts	
1878	Armstrong, UK	Sir William Armstrong, solicitor, engineer and industrial magnate, Newcastle-upon-Tyne with a passion for water-power, as opposed to coal Armstrong installed a hydro-electric plant, at 'Cragside' his Northumberland home, using lake water and a 4.5 kW water-turbine driven DC dynamo to light lamps in the house. Worked with local colleagues William Siemens and Joseph Swan	
1886	Tesla, USA,	First polyphase AC induction motor/generator developed	
	Dolivo-Dobrovolsky, Russia, Germany and Switzerland	First three phase AC induction motor developed, working independently from Tesla Then, developing a three phase AC induction motor with a polyphase slip-ring rotor connection to resistors for starting control	
1887	Charles Parsons, Newcastle, UK	4 kW, steam turbine driving a 2-pole DC generator with commutator	
	Prof James Blyth, UK, Scotland, Royal College of Science, Glasgow, for electricity production and battery charging	10 kW, 7-blade VAWT, driving a DC dynamo. Believed to have had adjustable blades, which contemporary alternatives did not. This is a large turbine, note Mrs Blyth standing below the VAWT	

(Continues)

Appendix A.3 (*Continued*)

1891	Denmark, Poul la Cour developed a wind turbine for electricity production and battery charging	10 kW, 4-blade HAWT, fixed pitch wind turbine driving a DC dynamo Innovative aerodynamic system and electric utilisation for electrolysis and hydrogen storage, mitigating wind variability with 20 years life and high reliability This is a large turbine, note man standing below the HAWT	
1904–11	USA and Germany, Ward Leonard, Kramer, Schrage	Development of variable speed DC and then AC electric generators/motors using switching via rotating commutators	
1910–14	Competition for electric windmills, from internal combustion engines, following the development of diesel- and then petrol-engine driven generators and low cost of oil		
1914–18	First World War, reduction in oil supplies, 20–35 kW electric windmills being built		
1918 post-war	Wind-mill development languished	Small wind turbines proving less reliable for electricity production than diesel- or petrol-engine driven generators, but AC electrical grid connection becoming more widespread	
1920s	Germany, Betz	Following First World War, influence of aerodynamic knowledge from aircraft starts to affect wind turbine design due to wing and propeller developments	
1923–73	Various inventors and industrial companies	Development of valve (ignitron) and semiconductor (thyristor) switches to convert AC/DC and DC/AC, facilitating reversible power flow through AC generators	
1930s	Denmark, USA, Electric windmills become common on large farms	Cheap high-tensile steel, windmills on prefabricated open steel lattice towers	Decline of American multi-blade turbine concepts in favour of European 4, 3, 2 and even 1-blade concepts
1930s–40s	Europe and N America, national AC electrification using fossil-fired power stations feeding national grids, based on Ferranti and Tesla's ideas with Westinghouse's investments. Research programmes in Denmark, France, Germany and UK considered wind power to supplement the national grid		
1932–50	Nyquist, Bode and others	Development of the techniques necessary to control the power flow through reversible AC/DC and DC/AC power systems, essential to the development of variable pitch, variable speed wind turbines	

(Continues)

Appendix A.3 (Continued)

1940	Germany, Ventimotor company formed by Ulrich Hutter. Cf, earlier light weight Chinese turbines	W34 2-blade down-wind HAWT cap-mill. Lightweight, cost-effective, guyed tower and typical excellent German aerodynamic design. Test centre near Weimar developing wind turbines for the war effort. Hutter (1954). This is a large tur-bine, note the man climbing the HAWT	
1900	UK, Isle of Man Cf, 1943, China large Norias	Laxey Hydraulic Wheel, water-driven steel structure, driving pumps to draw water from a lead, copper, silver and zinc mine Note two people above the water wheel	
1930	China, Shang Hai (上海).	Beginning of the electric age in China. 20 MW, coal-fired power station, behind Power Company offices. Efficiency 10%–20%. IET Shanghai collection	

(Continues)

Appendix A.3 (Continued)

1943	China, Gansu Province (甘肃). Cf, 1993, 3 Gorges Hydro Water Turbine, 700 MW, 50 years later!	Historic 1 kW vertical axis hydraulic turbine, for seed-milling or pumping, efficiency <5%. Published with permission from the Needham Research Institute	
	China, Chengdu, Szechuan Province (四川成都).	Historic traditional river-powered, large, 3 wooden, high-lift water Norias, 1–2 kW. For irrigation, efficiency <5%. Published with permission from the Needham Research Institute	
	Cf, 1900, Laxey mine water pump. and 2021, 55 kW diesel-driven agricultural irrigation pump	2 historic traditional river-powered, large quantity, wooden, high-lift water Norias, 5 kW. For irrigating large fields. Published with permission from the Needham Research Institute	
1943	China, Nei Chiang, Szechuan Province (四川内江). Cf, 55 kW diesel-driven agricultural irrigation pump, 2021.	An historic traditional low output wooden water raising pump. Pedal-driven, human-powered square-pallet, chain clack conveyor for irrigating small rice paddy fields. Published with permission from the Needham Research Institute	

(Continues)

Appendix A.3 (Continued)

	China, Chongqing, Szechuan Province (四川重庆). Cf: 1992 modern thyristor AC to DC link technology Gezhouba 2022	2×100 kW synchronous AC to DC rotary converters at an electro-chemical plant. Efficiency 60%–70%. Published with permission from the Needham Research Institute	
1944	China, Showing Chinese Industrial Revolution, following European and US with similar technologies.	Smelting iron in a cupola furnace at the National Resources Commission Machine Works. Lanzhou, Gansu (兰川 甘肃). Published with permission from the Needham Research Institute	
	Cf Abraham Darby's furnace, UK, 1790	Steel production in an underground factory built into the mountain at the Central Machine Works. Kunming, Yunnan (昆明 云南). Published with permission from the Needham Research Institute	
1944	China, Chonqing, Szechuan (四川重庆).	Coal waggons being loaded at the pithead, Tienfu (天府) coal mine, near the Jialing River (嘉陵江). Published with permission from the Needham Research Institute	

(Continues)

Appendix A.3 (*Continued*)

	Showing Chinese Industrial Revolution, following European and US Industrial Revolutions with similar technologies	Early steam locomotive for hauling the coal waggons on narrow gauge tracks to the Jialing River (嘉陵江). Published with permission from the Needham Research Institute	
		Factory buildings on a cliff at the end of the railway from the Coal Mine, above the Jialing River. Published with permission from the Needham Research Institute	
		A 10-oared traditional barge, hauling coal from a mine at Tianfu (天府) down to ChongQing (四川重庆) along the Jialing River (嘉陵江). Published with permission from the Needham Research Institute	

(Continues)

Appendix A.3 (*Continued*)

1947	USA, Smith-Putnam, world's first megawatt-size wind turbine connected to electrical distribution system Grandpa's Knob, Castleton, Vermont. Considered the grandfather of modern electric wind turbines	1.25 MW, D 57 m, 40 m high, 2-blade downwind HAWT, geared, constant speed, full-span pitch control, stall-regulated, synchronous machine. Sophisticated, modern, AC grid-connected wind turbine. Design by P C Putnam [8], including aerodynamicist von Karman, MIT. Manufactured by S M Smith	 *The turbine completed on Grandpa's Knob*
1950–70	Various inventors and industrial companies	Development of new types of power semiconductor switches, the thyristor, transistor, IGBT	
1957 onwards	Number of inventors	Back-to-back reversible DC and AC drives using thyristor or other semiconductor bridges with forced static commutation, switched twice per cycle. Static commutator eventually renders rotating commutators obsolete, developments essential for variable speed electric wind turbines using AC generators and converters and with electric pitch motors	
1966	France, La Rance tidal barrage	First tidal barrage with a capacity of 240 MW and a capacity factor of 24%	
1972	International oil crisis, triggered by Yom Kippur war, a renaissance of wind power, see Gipe (2016) considering both small and large wind turbines		
1975	USA	General Electric invention of new fast-acting semiconductor devices	

(Continues)

Appendix A.3 (Continued)

1975–85	USA and Europe	Aerodynamic and material development of blade aero-foils, air-brakes, pitch mechanisms and blade materials	Evolving from steel and aluminium to composite glass-fibre and innovative balsa and bamboo wood structures
1976–81	USA and Western Europe, modern small-scale wind turbines	1–10 kW, VAWT and HAWT	Small-scale inheritors of the 1930s small US and Danish turbines. The market was still uncertain because the technology is still unresolved
1979	Denmark, Nibe, 2 experimental machines erected, incarnation of the modern wind turbine. One with pitch control and one without. If Smith-Putnam the father of the modern electric wind turbine, these 2 were his strongest children	200 kW, D 24 m, 3-blade upwind HAWT, geared-drive, fixed speed, stall-regulated with aerodynamic rotor blade tip brakes 630 kW, D 40m, 3-blade upwind HAWT, geared-drive, full-span pitch control, fixed speed, stall-regulated	 

(Continues)

Appendix A.3 (*Continued*)

1980s	Greece, Island of Crete. Cf, Earlier Chinese wind turbines	Current example of historical HAWTs in rural Europe. Traditional HAWT for water pumping on Lassithi plateau, with luffable sails	
	Germany, Bottrop, on the Rhine-Herne Canal	Waste Water Treatment Plant (WWTP) pumping station using 3 parallel, electrically-driven, Archimedes Screw water lift pumps, serving 1,350,000 people	
1980s	USA, Western Europe and Japan	Replacement of variable speed drive bridge thyristors by faster, cheaper switches, using insulated gate bipolar transistors (IGBT), offering faster switching speeds and better waveform management. Development of PWM IGBT voltage source converters (VSC) drives with early controllers, static commutators enabling simple 1, 2 and 4-quadrant control of wind turbine AC and other variable speed renewable generators, rendering earlier rotating commutators obsolete	
	Netherlands	Development of modern 300 kW geared-drive, 3-blade, stall-regulated, fixed speed wind turbines	
	USA, Great California Wind Rush, following oil price rises	Large numbers of wind turbines $\leq$ 100 kW, mostly HAWT but some VAWT. Very poor reliability from many different designs but some did survive	
1993	China, Hubei Province (湖北). Cf, Gansu 1 kW Water Turbine, 1943. Cf, France-Spain IGBT HVDC Link 2022	Three Gorges Dam Power Station, SanXiaDaBa (三峡大坝). Includes 32× 700 MW, vertical axis hydraulic turbines driving synchronous AC generators, installed in the dam. Efficiency 95%.	

(Continues)

Appendix A.3 (Continued)

		HVDC substation using thyristors feeding 1200 MW, 500 kV DC from Three Gorges Dam GeZhouBa (葛洲坝) to ShangHai (上海) Converter station efficiency 90%	
2003	UK	Marine Current Turbines Sea Flow, 300 kW. Tested off the Devon Coast near Lynmouth, subsequently withdrawn from service	
2008	UK, Northern Ireland	Marine Current Turbines SeaGen twin tidal turbines, 1.2 MW. Tested in Strangford Loch, subsequently withdrawn from service	

(Continues)

Appendix A.3 (Continued)

2008	China, DongHai DaQiao (东海大桥), ShangHai (上海). Wind turbines mounted on concrete gravity caissons. Note the low top-head mass and slender steel tower	Installation of a 3 MW offshore 3-blade upwind HAWT, pitch-regulated, Sinovel SL3000/90, geared drive, fully-rated grid-connected IGBT variable speed converter	
2014	Germany, BARD Off-shore Wind Project	400 MW off-shore wind farm, 80 5 MW 3-blade, upwind HAWT, geared drive, doubly-fed wound asynchronous generator, partially-rated, grid-connected, IGBT variable speed converter. Mounted on 3-leg steel jackets, for deep water, rather than, more common earlier steel mono-piles	
		Wind turbines with step-up transformers connected by 36 kV AC collector cables to an off-shore sub- and IGBT HVDC converter-station. This inter-connects to shore from the HVDC converter-stations by 90 km, 170 kV DC cable links, enabling economic long-distance power transfer to a shore-based sub-station, where it is converted to AC and stepped-up to 380 kV	 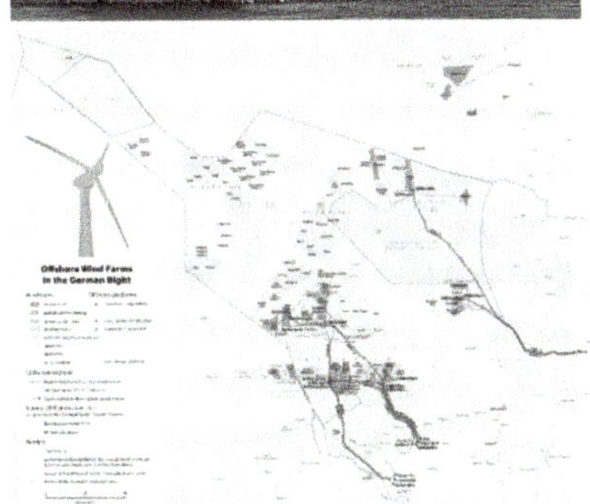

(Continues)

Appendix A.3 (*Continued*)

2014	UK, Orkney, EMEC test site	Pelamis P2, 750 kW wave device under test. Subsequently withdrawn from service	
	UK, Cragside, Northumberland	Pioneer of clean energy, Sir William Armstrong, his restored *Cragside* home with a modern, water- powered Archimedes screw driving an 18.5 kW AC induction generator, supplying electricity, replacing Armstrong's 1870 installation. He also founded Armstrong College in Newcastle, subsequently King's College, Durham University, becoming in 1964 Newcastle University	 
2018	Nigeria	Low power modern electrical networks. Using photovoltaics for low power lighting, computing and small battery storage. Efficiency 80%	

(Continues)

Appendix A.3 (*Continued*)

2020	Pakistan, Bin Qasium II power station	Combined cycle gas-fired power station, 573 MW. Efficiency 55%. Behind, a less powerful coal-fired power station	
2021	UK Cf, Gansu (甘肃) 1 kW water turbine pump, 1943	Modern 55 kW diesel-driven agricultural irrigation turbine pump. Efficiency 80%	
2021	UK, Scotland Cf, 750 kW Pelamis wave (2014), 1.2 MW Seagen tidal (2008) and 300 kW SeaFlow tidal (2003)	Orbital Marine Power's O2, 2 MW twin tidal turbine. Based upon experience from: • Pelamis wave device; • SeaFlow and SeaGen tidal devices	
2022	France-Spain Cf, GeZhouBa (葛洲坝)-ShangHai (上海) Thyristor HVDC Link, 1993	IGBT HVDC substation feeding DC power, 2,000 MW at 320 kV, between France and Spain, to balance constant nuclear against variable renewable resources. Converter station efficiency c90%–95%	

Appendix B

Two energy poems

Caliban's speech from The Tempest, William Shakespeare

Be not afeard; the isle is full of noises,
Sounds, and sweet airs, that give delight and hurt not.
Sometimes a thousand twangling instruments
Will hum about mine ears; and sometime voices
That, if I then had waked after long sleep,
Will make me sleep again; and then in dreaming,
The clouds methought would open, and show riches
Ready to drop upon me, that when I awaked,
I cried to dream again!

Isambard Kingdom Brunel declaiming at the 2012 London Olympics opening ceremony
played by Kenneth Branagh,
followed by the song 'Caliban's Dream', sung by Alex Trimble & Underworld

Doggerel

In windswept fields ere dreams took flight, a vision's born shrouded from light,
A friend an engineer by trade, talked of winds that could drive a blade.
His words a beacon in a sceptic's haze, faith in wind became a steadfast blaze,
Uncertain friends questioned his claim, but he held true ignoring disdain.
Now in that realm where progress resides, a triumph unfolds as truth abides,
News arrives with gusts of delight, half of Albion's electricity takes flight.
Wind turbines turn a majestic ballet, harnessing currents that networks convey,
Midst that doubt an engineer stands, champion of foresight in proud command.
For he looked beyond doubt that crept, embracing a future where energy swept,
While turbines turn on a breeze's wraith, a testament to vision and faith.
Let this ode be a heartfelt cheer, to our windy pioneer,
May that legacy cast a radiant light, guiding us to a future bright!
Anon (2023)

with thanks to Joe Ruston and homage to Professor James Blyth (1887)

Connectivity end-piece, courtesy Michael Carmichael

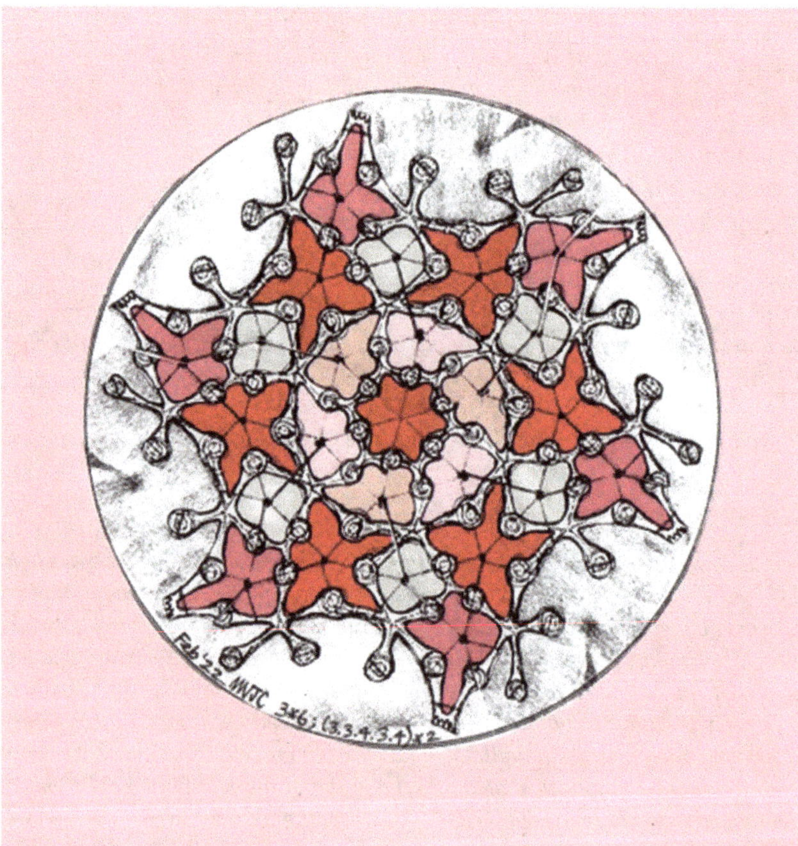

Index

www.ingramcontent.com/pod-product-compliance
Lightning Source LLC
La Vergne TN
LVHW011836200926
848484LV00002B/363